GOD DISCOVERY

HOW SCIENCE AND LOGIC REVEAL THE EXISTENCE OF GOD

H. KEVIN SMITH, PH.D.

God Discovery
by H. Kevin Smith
Copyright © 2021 H. Kevin Smith

All rights reserved. This book is protected under the copyright laws of the United States of America. This book may not be copied or reprinted for commercial gain or profit.

Scripture quotations marked NIV are taken from THE HOLY BIBLE: New International Version ©1978 by the New York International Bible Society, used by permission of Zondervan Bible Publishers. All rights reserved.

ISBN 978-1-63360-170-3

For Worldwide Distribution
Printed in the U.S.A.

Urban Press
P.O. Box 8881
Pittsburgh, PA 15221-0881
412.646.2780

Preface ... vi

Section One
My Background

Chapter One - My Life's Journey ... 3
Chapter Two - Switching to Chickens ... 13
Chapter Three - Another Change in Focus ... 21
Chapter Four - Academics ... 29
Chapter Five - Spiritual Growth ... 34

Section Two
God and Science

Chapter Six - Science Points to God the Creator ... 43
Chapter Seven - Evolution: An Implausible Theory ... 50
Chapter Eight - Challenging Chemocal Evolution ... 59
Chapter Nine - A Tapestry ... 70
Chapter Ten - The Human Body ... 84

Section Three
The Bible and Religion

Chapter Eleven - Science and the Bible ... 95
Chapter Twelve - Archaeology and the Bible ... 102
Chapter Thirteen - Textual Integrity of the Bible ... 107
Chapter Fourteen - Prophecies in the Bible ... 116
Chapter Fifteen - What About Other Religions? ... 123
Chapter Sixteen - The Journey ... 127
Chapter Seventeen - Heaven and Hell ... 136

Conclusion
Your Journey ... 145

Appendix One ... 153
Appendix Two ... 157
Sources and References ... 160
About the Author ... 163

Dedication

First, I would like to dedicate this book to my parents, Jewette Grant Smith II and Elizabeth Smith, who through love and a strong set of family, moral and work values set me on a firm foundation for my life. Second, for my academic foundation I dedicate it to my teachers and mentors, in particular Drs. Edward Wallace and Raymond Craig for their care and tutelage. Finally, the book is dedicated to my God and Lord, Jesus Christ, who has provided for and guided me every step of my life journey.

Preface

This book was written after many years of study, both in the scientific and biblical realms. For many, these two areas appear to be mutually exclusive without any common ground and diametrically opposed to each other. As a result of years of accumulated knowledge indicating that science and the Bible are actually compatible, I have offered a number of thirteen-week Sunday School adult classes. The information I presented almost always resulted in great interest, an increased enthusiasm for the listeners' faith, and the removal of a cultural curtain that had inadvertently promoted a view that science could not be in harmony with religion and would never meet.

My experience could not be further from this popular point of view of their dichotomy that even seemed to have infiltrated religious and Christian circles. The reality of their congruence was impressed upon me recently while having a discussion with a Christian I was hosting who was from New Zealand. He asked me in essence how I could sustain a deep faith in the God of the Bible while still firmly holding to my doctoral studies and research in the area of physical chemistry. I responded to his amazement by stating that science and logic have actually laid a *stronger* foundation for my faith.

Even today, I am amazed at the additional scientific knowledge that without a doubt points to God as the Creator. All of nature has God's fingerprints on it. What is even more incredible is that numerous Christian scientists and professors choose to disregard these obvious connections of science with God. Either they do so for fear of losing their "academic" credibility or wanting to separate their scientific profession from their Christian faith.

In an attempt to provide an opportunity for people outside of a church setting to see highlights from this Sunday School class information, I also developed a presentation offered during a three-course meal. This informal presentation, which many times churches have sponsored, I call the "God Discovery Banquet." More information on this program is provided at www.

GodDiscoveryBanquet.com. The presentation has always generated a great deal of interest and those who attend hear and see an overwhelming amount of evidence for not only the existence of God but also more specifically the God described in the Christian Bible.

I have never received any negative feedback from those who have attended a God Discovery Banquet. No one from the field of science or engineering has refuted what I present. Some even comment that they were "blown away," those remarks sometimes coming from individuals not familiar with the Bible once they have heard the science, archeology, prophecies, and other information given during the presentation.

After years of study to gather an overwhelming amount of information pointing to the existence of God the Creator, I now desire to share this information, in a wholesome and God-honoring way, with others who have an open mind. Actually, this way of thinking is not new but was held by many of the time-honored pioneers of the sciences, which is a topic I touch on later in the book. This is my motivation to write this book. My hope is that through sharing what I have learned, I will stimulate you to seriously reflect upon what I present. I pray you may be curious as to why I would take such a strong stand and therefore sincerely examine and not dismiss the science I will present. The same is true for those who may ridicule the idea of God being the Creator of all. Please consider my challenge for you to examine what I herewith present.

In the first few chapters, I share my life's journey and early experiences, which include my life growing up on a farm in rural western Pennsylvania. Many people have found those stories fascinating and amusing. I will also share my journeys both in my academic and religious life. We all have different and unique life journeys to share, and sharing mine will help you understand who I am. I will make a connection between my growing up and my current understanding of science. I will also show how the Christian faith is unique, standing apart from all other faiths, with the main foundation of the Christian faith being the Bible. Finally, I will present evidence showing the Bible to be the word of God.

I would like to thank my dear wife, Janet, for all of her loving support during my life's journey and our life together since 1983. She has helped me deepen my relationship with God and my work in the areas of science and technology. It is my hope that after reading this book you will gain a deeper appreciation of who God is and then be motivated to continue your life's journey with a sincere and stronger intent to know God by exploring and developing a closer relationship with Him. In the end, may God ultimately receive the glory.

                        Kevin Smith
                        August 2021
                    West Leechburg, PA

SECTION ONE

My Background

Chapter One

My Life's Journey

Why do you believe what you believe? Many of us have beliefs that have their roots in our childhood and are comprised of how we were raised, the experiences we had, and the memorable impressions we can recall. It is human nature to avoid many of the important, life-direction questions and instead continue on our day-to-day routines of life with a false sense of security in a belief that our tomorrows will always be there.

However, the hard fact is that as time marches on and our finite existence will become more and more certain. Sometimes a life-changing event such as the death of a loved one, loss of a job, a life out of control from addiction, or some other earth-shaking event will cause us to think and ask questions regarding what is truly significant, what is eternal and certain. We may even go so deep as to consider if there some common denominator for all life that transcends our finitude and personal life philosophy.

For me, the foundation for my belief(s) began during my childhood, though at the time I did not fully realize this. Only in my latter years as I looked back could I see a rational and sense of logic with a simpler approach in regard to the existence of God being the foundational building blocks for my faith. This even included the day-to-day growing-up experiences and patterns of my life. I will now proceed to write about my experience growing up on a farm, discovering my interest and talent in the sciences, and coming to faith—all while living in rural America. For some, this may not be of interest or the reason for reading this book, in which case feel free to skip ahead to the scientific and apologetic chapters starting in chapter 6. For others, what comes next may actually entertain, at times amuse, and even give you a glimpse of farm life and its multi-faceted activities.

THE HARVEST

I was raised on a small 100-acre farm in southwestern Pennsylvania. I enjoyed a very good childhood, the youngest of three sons who were raised by a loving mother and father steeped in a traditional rural culture. We were not wealthy but had enough and were satisfied with what we had. We never went on a vacation and only had one car which we always purchased used. All the farm equipment was older but well taken care of. It was a very satisfying and relatively enjoyable childhood filled with good memories and experiences, not to mention the responsibilities and chores associated with that kind of life where family mattered. We all knew this not by merely verbally saying we loved each other, but also by what we experienced and expressed in our day-to-day life.

Dad began his farming business by raising 30 to 40 beef cattle and selling them at auction. Many of the work activities associated with this farming business included planting barley, oats, and corn in the Spring. A significant amount of work needed to be done to the ground before one seed was planted. Plowing was first done to turn over about ten inches of soil, using a two-bottom plow with two curved, elongated pieces of iron that cut and flipped over the soil when the tractor pulled it. The end result was a rough-and-bumpy ground surface that was worked until it was a smooth surface with looser soil after a cultipacker and harrow were also pulled over it by the tractor.

A grain drill and corn planter, also pulled by the tractor, would automatically plant the seeds into the prepared soil along with fertilizer in granular form. The planters were fascinating to me. They were both powered by the planter wheels and wheel axle that had turning gears as the planter was pulled across ground. The gears would activate various mechanisms that would drop each seed into the newly prepared ground. For corn, once the seeds sprouted, the prolific weeds that would also sprout needed to be removed by a cultivator attached to the frame of the tractor that was driven down between the corn rows.

The corn itself was planted in rows about two feet apart. The cultivator's multiple spade-like prongs would dig into the dirt, uprooting and killing the unwanted weeds between the corn rows. Weeds left among the corn rows would be killed by a special herbicide that did not affect the corn. A 55-gallon drum was mounted on the back of the tractor, along with a sprayer pump connected by hoses to spray nozzles on booms extending out from each side that would spray the herbicide on the corn rows as the tractor was driven between the rows.

Harvesting was the next phase for the barley, oats and corn crops. Harvesting the barley and oats crops was accomplished in the summer by hiring a local farmer to cut the grain with his combine, while we used our tractor and wagon to transport the barley and oats to the barn granary. At this time, I was in grade school and was more of a helper and laborer and really did not operate any of the machinery. During the harvest, I was fascinated with Bill, the hired farmer, as he drove his tractor and combine and interacted with my dad.

They were good friends and Bill was a salt-of-the-earth kind of guy. I can still hear his loud conversations with Dad about the weather, the crop harvesting conditions, and current issues of farming in general. He was thinner, with a good farmer's tan that had produced a weathered face. My dad would also hire Bill to buy younger stock beef cattle at an auction house that we would fatten up over a few years to eventually sell at auction—again with Bill's help.

Bill had a daughter, Susan who was my age. On a few occasions, I accompanied them to the auction and watched with eager curiosity to see the animals auctioned one at a time. The auctioneer would rapidly shout out the bidding for each animal in the typical rhythmic and rapid auctioneer style in their almost song-like fashion, making a somewhat boring series of animal sales seem like a concert or sporting event.

Another thing that intrigued me was watching Bill make his own cigarettes while he talked with Dad. He would take a single sheet of paper from his packet of cigarette papers, skillfully hold it in the shape of an open tube situated between his fingers on one hand, while taking a small pouch of tobacco with a draw string.. He'd pull the string with the other hand and carefully sprinkle the tobacco onto the cigarette paper. Next, he carefully rolled the cigarette paper to keep the tobacco within the cigarette, finally licking the edge of the paper so it would stick to the rolled side of the cigarette—and not drop a speck of precious tobacco. He would then pull closed the draw string on the pouch, put it in his pocket, get a match, and light his newly formed, self-made cigarette. Bill did all this while simultaneously talking with my father without missing a beat.

My earliest memory of this harvesting activity was Bill's grain combine harvester. It was pulled by a tractor, like all our equipment, but the power to move the cutting blades and threshing items was provided by an on-board motor. Someone had to ride on it to catch the grain in sacks, which were then closed shut with short pieces of bailor twine. Once a few sacks were filled, the person would slide the filled sacks down a small shoot the short distance to the ground. Filling the sacks, tying them closed, and releasing them onto the ground occurred while the combine continually moved along the field cutting and threshing the grain.

I would ride along the harvested field in a wagon pulled by, our tractor, stopping at each point where the filled sacks were ready for pick up. Once the wagon was full, the heavy load was pulled back to the barn to be unloaded. The sacks were carried to the wooden granary bins inside the barn and emptied. The empty sacks were then delivered back to the person working on the combine. When I was young, I was not able to help lift the sacks of grain, which were probably 70 to 80 pounds, but I was able to help with the empty sacks and twine. I felt so mature and grown up when I was able to help.

The Chaff

Many times, Bill had difficulties with the combine motor, first starting it and then trying to keep it running. When harvesting season was upon us one year, we heard the exciting news that Bill had finally taken the plunge and purchased a new combine. This was big news in our world. Little did we know that the whole process of collecting the filled sacks was going to change. The combine he bought did not have a motor but had a connection for power to the power take-off from the tractor. That did not really affect us other than the harvest time was not interrupted with combine motor break downs.

The main difference for us was that the threshed grain ended up in a bulk bin on the combine which was emptied by an auger shoot. The shoot was a tube extending at a 45-degree angle from the bottom of the bin, with a large screw-type auger that when turned would move the grain up the pipe and into another mobile grain bin for transporting back to the barn. The problem was that we did not have a mobile bin. Instead, we used the burlap sacks to collect the grain from the shoot, tie off the sacks, and stack them in the wagon. It quickly became obvious that we had taken for granted the person on the old combine system who had previously filled the sacks.

The first few attempts at using mechanical devices to hold the sack for filling were failures because the force of the grain hitting the sack dislodged the sack especially if it was not perfectly in place. The only solution was to manually hold each sack on the wagon floor, which required two people facing each other and using their four hands to keep the sack open to catch almost all of the grain exiting the auger shoot three or four feet above the sack opening. As the grain would fall from the end of the shoot, which was slightly above our heads, into the sack, the two holders would work as a team moving the sack opening to catch the grain while adjusting the sack position to account for any direction changes due to the grain flow speed and wind.

The wind was what made this sacking operation most challenging, not because of the slight difference in direction of the grain it would create, but due to the dust and chaff that the wind would remove from the grain as it left the shoot opening. When the wind changed directions, it always seemed to blow directly into the face of one or both of the holders during the course of emptying the combine bin. By this time, I was old enough to hold one side of the grain sack, getting covered with dust and chaff in the process. The grain harvest all happened during summer, many times when the weather was hot and humid, which meant we were soaked with perspiration.

Sweat proved to be an excellent means through which the chaff and dust would stick to our arms, face, and neck. Because of this experience, I can relate to the threshing floors mentioned in the Bible where the thresher would toss the grain into the air and have the wind remove the lighter chaff. Despite the dirt and hard work, we would always successfully harvest the grain. This imparted to me an important life lesson that perseverance during hard and dirty work would lead to satisfaction at the end of the day once the job was finished and done well.

Corn

The corn harvest was a completely different story because it entailed a team partnership between my uncle, another long-time friend who also had a farm, and my dad. The three farmers had all put their money together to purchase a corn picker. Over a few weeks, they would join forces to harvest the corn as a team from their farms. My uncle would use his tractor to pull the corn picker, while my dad and his friend would bring their tractors with wagons that had two-foot-high sides on the perimeter of each wagon to hold the harvested corn cobs.

The corn picker machine would separate the corn ears with their hard dried kernels on the cob from the dry corn stock as it was pulled through the corn row. A pair of spinning parallel rollers would grab the husks and remove them from the corn cob, leaving behind the cob with kernels attached. A shoot that was elevated over the wagon attached to and pulled behind the picker shoot would take the harvested cobs and transport them up the shoot, dropping the cobs into the wagon. While riding the wagon my task would be to retrieve a few of the cobs that did not completely have the husk removed by the rollers and remove the husk by hand.

Once the wagon was full, a tractor would take it to the corn crib where the cobs were shoveled into an elevator that dumped them into the corn crib building. Our particular corn crib was actually an open building where the three walls served as the cribs. The cribs were constructed with double walls with parallel vertical boards with an inch gap and about five feet between the inner and outer double wall. The corn cobs were then dumped between the inner and outer walls. The gaps in the boards allowed air to pass through the corn to accelerate its drying.

I played in the crib when it was half empty, climbing through a shoot door used to remove and sack the corn, and going through the inner maze up and over the remaining corn cob piles. A peaked metal roof extended over the three walls, with an area inside being defined by the u-shaped walls used to house farm machinery. Later I found out Dad built the intricately designed corn crib before I was born. He designed it, used wood from trees that grew on the farm, and built it using only hand tools. Today, as I reflect on this crib and the many other things he built, I am amazed at what he could do and accomplish using only what he had at hand.

Another important summer activity was growing the grass fields that consisted of timothy, orchid grass, and alfalfa. These were cut with a sickle bar mower hooked to our ever-present tractor and dried over a few days. Then a rotating rake pulled by a tractor (what else?) rolled up the cut dried hay into a fluffy row. When the sun had sufficiently dried the hay, a baler would pick up the hay row with rotating teeth and then auger it into a horizontal steel shoot where the hay would be pressed into the familiar rectangular bale shape. The bales were held together by two parallel twine strings that were mechanically tied once the bale reached sufficient size. Each bale was around three to four feet long, two feet wide and 18 inches high, weighing about 50 pounds. We carried those bales by gripping the two twine strings that held the pressed hay together.

The tractor not only pulled the baler, which was powered by the power take-off of the tractor, but also pulled the hay wagon attached to the back of the bailor. Someone would drive the tractor while another would be on the wagon pulling up the hay from the bailer shoot with an elongated metal hook, grabbing the bail by its twine strings to stack it on the wagon. When I was younger, many times I would just ride the wagon. As I got older, I helped my dad use the hook to pull up the bail then sliding it back to him where he would stack it. Once the wagon was full, we would disconnect the tractor from the baler and connect the wagon to the tractor to take it to the barn. There we would unload the bails and stack them on both sides of the barn up to the ceiling, which was probably about 50 or 60 feet high.

My brother would unload the wagon, placing the bales onto an elevator, which was like a conveyor belt that carried the bales up to us. Then Dad and I would retrieve each bale to build a layer that would eventually become the floor for the next layer. Once the wagon was empty, we would climb down from the haymow using the ladder built into the vertical beams holding up the roof. The rungs consisted of 1.5-inch in diameter and two feet long pegs inserted every foot into the vertical beam. A smaller wooden beam placed parallel to the bigger beam accepted the other end of the rung into a bored-out hole.

After we unloaded wagon upon wagon filled with hay, the height of the haymow got bigger and bigger. The gradual increase of height helped me gradually get acclimated to the height I had to climb up the ladder until I reached the top of the beam probably 30 to 40 feet up. However, the height did not stop there since we increased the height up to the roof that sloped into and up to the roof peak, while keeping a small area near the ladder at the same level so we could get to the ladder. This meant we created sort of a double layer, one at the ladder where one of us would retrieve a bail and throw it to the upper layer where the second person would build additional layers that extended to the roof.

One noticeable difference as we got to this upper level was the increase in temperature, analogous to what one feels when going into an attic in the middle of summer. I always looked forward to putting the last bails near the peak since I would be high enough to go to the wall at the peak and look out a small window, imagining what birds would see from the vantage point. With the barn already on top of a small hill, the view seemed even more spectacular. Once completed, half of the barn was filled. The next wagons of hay were then used to build the haymow on the other side of the barn. In between the two haymows, we stored farm machinery such as the hay bailer, hay wagon, and grain and corn planters.

During the year, particularly in the late fall and winter months, the hay bales would be retrieved from the haymows, maybe twelve at a time, to feed the beef cattle at the lower ground level of the barn. The pasture was not adequate to feed the beef cattle during these months, so their diet was supplemented with hay and ground feed from the corn, oats, and barley we had harvested. I would climb up to the top of the haymow, carry a bale to the edge, and drop it over the side. This required some finesse since if they landed on the floor on the end of the bale, they would often break one or both of the twine strands holding the bale together, creating a mess of loose hay that would need to be pushed down the steps and gathered with a pitchfork. The higher up they were, the more careful I had to be with each launch.

It did not take long to develop the skill of simply dropping the bale so it would land level with the floor creating a loud thump. After climbing back down the ladder once the desired number of hay bales were dropped, the bales were carried over and slid down the steep steps to the feed room. We leaned the bales vertically so they stood against the back-stone wall, the twine cut with a knife, and the hay fed into a wooden manger above the feed trough where the cattle could pull out bunches of it to eat.

When I was in first grade, Dad was only raising beef cattle, which entailed all the growing and harvesting of the hay, grain, and corn for feed. After winter, we had to remove with a pitchfork the manure that accumulated in the barn and place it into a special wagon called a manure spreader. Like other equipment, the wheels powered it as it was pulled across the ground. Not only would the power be used to move a conveyor-type system to slowly push the pile of manure to the back, but the power from the wheels would also be used to spin teeth and paddles in the back that would randomly and more evenly spread the manure through the air on to the ground.

Needless to say, this vibrant spreading action had the side effect of injecting the aroma of manure in the air. It amazed me that Dad would manually remove the two feet of manure by hand using a pitchfork and the manure spreader. Since this was a spring job, Dad would do this job mostly by himself since we were in school—and for that I was grateful.

Chapter Two

Switching to Chickens

Dad focused on buying young beef cattle, feeding them for a couple of years and watching them grow bigger until his friend Bill took them to an auction to be sold for eventual slaughter. Dad did that until the local feed mill convinced him to raise one-day-old chicken peeps until they were twenty weeks old when they could be sold to other farmers who would use them for layers to sell their eggs. We would get 10,000 peeps twice a year that were then raised in a block building (Figure 1) we had to build that was the length of a football field (300 feet by 60 feet). This also occurred while I was in elementary school, so I was too young to appreciate what it took for Dad to pull this off.

He had to plan the construction of the building and purchase all the equipment (like brooders, feeders, and waterers), assemble the equipment, build wooden roosts, and get ready for 10,000 peeps to arrive in four dozen cardboard cartons. I waited with great anticipation for the first arrival of those peeps, having prepared for their arrival by setting up the brooders. Brooders are a metal hexagon dome, six feet in diameter, with a propane gas heater to provide heat and simulate a mother hen's warmth for 500 chicks. Each brooder came unassembled in a big flat cardboard box so we had to put them together, which was like working with a big erector set as we bolted together the six faces and the top with the heating element to create the final dome shape.

Figure 1. Picture of a little more than half of the building built to raise chickens, later converted to a sheep barn for 200 ewes.

I used my small fingers to take the small bolts, washers, lock washers, and nuts to assemble dome faces, attaching the six red legs to the dome and canvas curtains that were fixed between the legs to help keep heat in the dome. They also served as a cloth door for the peeps to enter and leave the heated dome. We took bales of oat and barley straw and spread it around the entire floor, except for the area where the brooders sat. That area was reserved for a much finer type of sugar cane straw composed of short pieces so when spread they would create a short carpet-like layer more appropriate for tiny chicks. The fluffy oat and barley straw created a six-inch layer in which the chicks would get lost.

The new fresh smell of the dispersed bedding permeated the barn as we prepared for the new arrivals. There was also a one-foot-tall cardboard fence that came in a roll with an aluminum foil face on one side. This fence kept the chicks close to the brooders, positioned at the boundary of finer sugar cane and the fluffier barley and/or oat straw. It was unrolled around each brooder with the aluminized side facing in and held together with clothes pins where the beginning and end met. I can still hear the noise hundreds of chicks made as they pecked at their image which they saw in the aluminized foil face of the cardboard fence.

The first few weeks of raising chicks required the most intensive labor. We literally had to put all 10,000 chickens to bed during week one because their natural tendency was to lay down next to each other. The problem was that they wanted to lay down against something hard like the border fence which I suppose was a substitute for feeling the security of the mother hen. However, we needed them to be close to the warm brooder which simulated the mother hen's heat. If they laid against the fence, they would just keep piling on top of one another to get more warmth until they would smother one another. We could not allow them to fall asleep along the fence.

The solution was to walk around each brooder/cardboard fence enclosure and yell to scare the chicks so they would run under the brooder. All five of us, including my mother, had to work together to train them to sleep under the brooder. The first few nights were particularly difficult. We would no sooner get the chicks under one brooder and move on to the next when they would move out and back to the fence. We would be running around and after a while we were exhausted and the chicks were too, dropping down to sleep where we didn't want them to be. We would then have to get down on our knees and literally scoop up the sleeping chicks, shuffling them back under the brooder. Fortunately, the chicks learned to sleep under the brooder, meaning each night we needed to move fewer chicks under there ourselves. After ten days, we would simply walk up and down the barn once to ensure all chicks were laying inside and around each brooder.

Feeding the Chicks

During the first couple of weeks, we manually fed and watered the chicks. Each brooder would have four or five cardboard flats (we assembled those) and five inverted one-gallon water bottles with screw-on water trough lids. We needed to fill those feed flats and water bottles one or two times a day, not an easy task considering the number of times the five-gallon water pale we used to fill those jars needed to be refilled using a spicket located at the middle of our football-field-sized barn. You would be surprised how much water and feed 10,000 chicks need.

As the chicks grew into their rather ugly, scraggly adolescence, they would jump over the cardboard fence. That was our signal to lower the metal trough feeder that stretched half the length, then width, then half the length of the building and back. We lowered the feeder with a hand crank that unwound a cable strung through pulleys on the ceiling and then attached to the feeder troughs. At the end of 20-week chicken-raising cycle, the cable would be wound back onto the spindle with the hand crank, thus raising the feeder trough to the ceiling to allow for cleaning out the barn of chicken manure.

After the feeder trough was connected to a bin in the feed room at the middle of the building, we could then automatically feed the 10,000 chickens by pulling the feed through the trough by a chain with paddles. We also set up portable automated watering troughs. This was a glorious event since it meant we no longer had to hand carry the hundreds of gallons of food and water for the chicks. There was still a brief transition period when we had to carry some feed and water to them before they learned how to access their new water and food sources.

During this transition period, we had to babysit, or should I say chick-sit, our chicks. The chicks were still small enough to jump into the trough and feed underneath the trough guard. This in itself would seem harmless except for their tendency to poop into the trough. However, when the chain was activated and moved to replenish the food pulling it though the trough from the feed bin, those adolescent chicks would ride the chain while eating. This ride could turn into a disaster for the chick when its claws would get caught in the chain or chain paddle. When that happened, they would frantically squawk until one of us could release them. If we were not there and/or could not reach them in time, they would be pulled through a corner gear sprocket and meet their demise. This sadly happened more than a few times.

The chickens sleeping arrangement also changed. Dad built dozens of chicken roosts, eight-by-four-foot structures with five two-by-two-inch boards stretched along their eight-foot length. The boards were one foot apart and about four feet off the ground. The boards were made so that at night, or even during the day, the chickens could jump up and sit on the board, all lined up in a row side by side. We had to train the chickens to sleep on these roosts at night, which meant more sessions of running, hollering, waving, and clapping our hands to scare them up and on to the roost. The act of roosting must have been more natural to them for we did not have to do this for very long. Once on the roost, it was only the matter of a week before they stayed there.

As the chickens quickly grew and ate tons of ground corn feed, it looked like every square inch was covered with live chickens, scampering and meandering throughout the building. It gave the appearance of a white carpet covering the entire floor. When walking through the barn, we had to exercise extreme care not to make any loud noises or quick moves, startling any of the chickens and causing them to fly. If that happened, it seemed like all 10,000 chickens would fly at once in random directions while flapping their wings and uttering squawking sounds, creating quite a panic not to mention the resulting massive dust cloud.

We quickly learned how to move through the barn with smooth motions as we performed our chores, like the weekly cleaning of the automatic watering troughs, the filling of grit hoppers, and opening and closing 40 or 50 windows on the 300-foot-long front wall. During this time the once fresh and clean straw would mix with the chicken droppings to create a fine, dirt-like floor surface. One side effect was the creation of ammonia gas that at times would give off a pungent odor throughout the barn. To alleviate this odor, we installed and evenly positioned a dozen large exhaust fans along the back wall. When the ammonia level was too high, we adjusted the fan timers to run more often to remove the ammonia-laden air.

Raising chickens for use in egg laying took a tremendous amount of organization, set up, and work. None of which was accomplished through random chaotic activity. This observation actually fits and is explained well by the second law of thermodynamics, a main topic that will be discussed in more depth later in the science section of the book. This law of science basically states all things naturally go towards disorder. To counteract the disorder, which is called entropy, ordered energy needs to be interjected. My brief description above of all the farm activities shows without our concerted effort, the chicken project would have degenerated into failure and the end of the chickens. The same is true for the universe in general. I mention this to hopefully whet your appetite for what is yet ahead and the main purpose of this book: to provide scientific evidence for the reality of God, the maker of heaven and earth.

Time to Go

As the chickens approached the twenty-week mark, their growth to adulthood was evident so they were ready to be sold to farmers to use as layers. For several weeks prior, some had started laying eggs so our walks through the barn to check on the chickens included looking for laid eggs and it was like an Easter egg hunt. It would not be uncommon to gather a few dozen eggs during a sweep of the barn. We would have so many eggs in that short time that eggs supplemented our cat's and dog's diets during this time. Any time friends stopped by, my parents would ask if they could take some eggs. After agreeing to take a few, they would be surprised when they saw the cardboard box filled with eggs that we handed them.

I never was privy to any of the actual sales transactions but was fully involved in catching and crating the 10,000 chickens over about a week. Catching the chickens would occur in the dead of night while the chickens were sleeping and roosting, making it fairly simple for a team of two to gently pick up two chickens at a time and place them in wooden cages or crates as they slept. For anyone outside the barn looking in the windows, it must have looked like some sort of clandestine operation with flashlights streaking back and forth in a pitch-black barn.

Since I was too young, I would not do the actual catching. Instead, I would man the crate door and flashlight in hand as my partner would scoop up a pair of chickens and place them in the crate as I opened and closed the crate door. Each pair would be placed through the door opening until each crate had 16 chickens. Once we reached the quota for the sturdy three-by-two-by-one crate, we each would grab one end and carry it together and stack them six to seven high next to one of the back doors. We then collected an empty crate, retraced our path to the roost, and filled the next crate.

Dad hired his friend Bill to transport the chickens to their new home and it was up to us to load them on the truck. Depending on the size and number of orders, we would fill 15-100 crates. I rarely went with the loaded truck to its destination but waited until the truck came back for the next load. If empty crates were available, we would fill them while we waited for the truck to return with empty crates. Once we saw the headlights of the truck returning an hour or two later, we would get ready to repeat the same chicken catching procedure.

Once all of the chickens were sold and the barn was empty, we took the last step in raising a 10,000-chicken batch, which was cleaning out all the chicken manure and disinfecting the barn for the next arrival. The feeders were raised, the watering troughs moved out of the way, and a small tractor with a high lift would load a manure spreader so we could spread the rich fertilizer on the farm's fields. Dad bought this particular tractor fitted with a front loader specifically for the purpose of cleaning out the chicken manure. This tractor did double duty to load the beef cattle manure as well, making that chore much easier.

Flat shovels would be used to scrape and shovel the powdery manure (sometimes twelve inches thick) away from the barn wall and into the high lift bucket or manure spreader. Once all the manure was removed, we would spray the hard-packed dirt floor, the roosts, and lower barn walls with a special solution of disinfectant. We used the sprayer attached to the tractor to spray the disinfectant in the barn. We again purchased sugar cane bales and used the barley and oat straw baled on the farm, distributing the baled materials the entire length of the barn.

The sugar cane bales were made up of crushed stocks from which the sugar had been removed. We cut the wire binding each of the sugar cane bales and then we spread the crushed the stocks under the suspended brooders. The cane stocks were compressed tightly together making it difficult to spread, meaning we needed to break it apart by hand. We covered the remainder of the barn with the barley and oat straw, scattering it with pitch forks. We lowered the brooders from the ceiling, put up the cardboard fence around each brooder, assembled the feeder flats and placed them within each cardboard fence, and distributed water bottles. We filled the water bottles and poured feed into the cardboard flats. After all that, we were ready for the arrival of the next 10.000 newborn batch of chicks delivered in a single shipment. The work on a farm is indeed never done.

Chapter Three

Another Change in Focus

We continued this cycle of raising chickens until my father had a life-threatening and life-changing health issue. We were a couple hours away from home at a homecoming event and football game where my brother had started his freshman year of college. During our time there, Dad experienced chest pains which prompted us to find a local doctor and have him examined. The doctor determined he had a heart-related event which he termed a heart spasm, meaning it was not a full-fledged heart attack. However, the event was serious enough that he directed us to take him to the local hospital for treatment and rest. I was in seventh grade and very concerned. My brother wanted to come home with us, but my parents convinced him to stay, assuring him we would be fine. My mother drove us home while I manned the map. We only made one wrong turn, resulting in us seeing some back roads we had never seen before.

While Dad recovered in the hospital for a couple of weeks, I was responsible for the 10,000 chickens who were older at that point and did not need our undivided attention like when they were young. Thankfully, our neighbor Joe who rented a small house adjacent to our beef cattle barn took on the responsibility of feeding and watching over the 30 head of cattle. I assumed responsibility for the chickens, filling the feed hopper and checking on them before school. My end-of-day and weekend work included cleaning the many automatically filled watering troughs, opening and closing the 40 or so windows as the weather dictated, filling grit hoppers, and periodic walkthroughs. For many years after this event, my mother often said that those few weeks were when I truly grew up as I took on those responsibilities.

When Dad returned home, he was restricted to bed rest for another month to six weeks so his heart could fully recover. This was really difficult for my father to do, but he did follow doctor's orders and eventually and gradually got back into the farm routine. After his return to farm work up until the time we sold the farm four years later, I worked much more closely with him, gaining more practical knowledge along with a deeper appreciation for him. I watched him much more carefully, insisting I do a lot of the heavy lifting work fearful of him having a heart attack.

Dad fully recovered, but other changes were on the horizon. We were about to hit the wall that many other large-scale chicken producers experience, and that wall was disease. Various chicken viruses began to plague the large chicken barn, resulting in having to inject certain medicines into the chicken's watering system. This represented a large expense but was also not fully effective. A local expert said installing a cement floor was a possible solution. After my father considered the large expense to cement the floor, which still may not solve the problem, he decided to get out of the chicken business.

The decision not to raise chickens resulted in another significant change in direction in farming. For as long as I could remember, Dad had a dream of raising sheep. He had a few old books on sheep which I assumed he had from his time at Penn State University from which he graduated in 1932. During a visit to Penn State a few years ago and long after my dad had passed away, my wife and I decided to visit the alumni office to see if we could find any records of his graduation. In our search, we found him in the 1932 yearbook where it stated he graduated with a degree in animal husbandry and was a member of the Block and Bridle Club, an agricultural group.

As a kid I looked at those sheep books out of curiosity, examining the different pictures of the various breeds of sheep.

Dad embarked on this dream of raising sheep long before the idea of raising chickens was hatched (pardon the pun). First, there was the building where he initially planned to house sheep. After plowing a field for planting a crop, stones would naturally be uncovered and come to the surface. After plowing, we had to break up the clods of dirt by pulling a cultipacker behind a tractor. If we came across rocks, we would stop and pick up the rocks, placing them on top of a shelf on the cultipacker. These rocks would then be piled at the sheep building site.

During some spare time, Dad would work on building the wall using these stones. He started with the back wall since it was against a small hill which meant he had to build it out a few feet wider to hold back the hill. He would make mortar using sand and a bag of mortar in a steel wheelbarrow with a steel wheel, adding water and then mixing it up. I still have his old wheelbarrow I sometimes use to mix mortar, fondly recalling those times Dad used it. It actually has some old crusty mortar on it—which probably came from him.

He would search through the pile of rocks for a desired shape and size to fit a spot on the next layer of this back wall. Needless to say, it took hundreds of stones pieced together into a sort of three-dimensional puzzle to build the wall. When I asked him about the rest of the barn, he would point out the location of barn doors and would walk through them, which told me he had them etched in his mind even though they were still invisible to me. After browsing through the sheep pictures in the book and seeing the beginning phases of construction on the sheep building, my interest was piqued and I eagerly anticipated the new experience of raising sheep.

This dream of raising sheep would not come to fruition until a number of years later and would not actually include this particular barn being built by hand at a slow pace. Once raising chickens became a priority, the partially constructed sheep barn instead became the newly-conceived hatchery building for chickens. Once we obtained the loan for launching the chicken business, construction of the hatchery took on a faster pace and focus. We also built a second and larger 300-foot building adjacent to the hatchery where the peeps were raised during their 20-week stay. When Dad made the decision to switch from chickens to sheep, he modified the larger 300-foot building for housing and raising sheep.

This required a major transformation to this building. The automatic chicken feeder was raised to the ceiling. The many chicken roosts Dad made were dismantled in order to reuse the wood and even the nails. Many of the nails needed straightened on a flat hard surface with a hammer to make them straight again. This act of nail straightening was typical of Dad's efforts to reduce costs by recycling nearly all of the chicken roosts. The recycled elements from the roosts were almost exclusively used to construct the sheep feed manger he constructed, dividing the 300-foot building in half lengthwise.

Posts were driven into the dirt floor eight feet apart. The two-by four-inch boards from the roost were used as the manger framing. The one-by-ten-inch boards from the sides of the roosts were fastened to the framing to form a feed trough that stretched the length of the barn. The two-inch-by-two-inch boards, the actual roost elements on which the chickens sat, were fastened vertically in front of the feed trough, parallel to each other and about six inches apart. These parallel positioned boards allowed the sheep to stick their head through to reach the feed trough.

After months of preparation, the barn was ready. Hay was bailed and taken to the barn, put through one of the windows on a simply built wooden shoot, and stacked on the side of the barn separated from the sheep side by the manger. In addition to the barn preparation, pastures were needed, not one or two but three, for the sheep had to be rotated every week. This would prevent the pastures from being over-grazed since sheep will literally eat the grass all the way to the ground. In addition, rotation of pastures would help break the cycle of parasitic worms the sheep can pick up while eating the grass. Each pasture was around ten acres.

I learned a great deal about how to put up a good fence during this time. Woven fence with a strand of barb wire added to the top and below on the ground were meant not only to keep the sheep within the boundaries of the pasture but also to keep out the occasional roaming dog or dogs that could run down and hurt the sheep. After the pastures were completed, I let our dog into the pasture to see if he could escape beyond the fence. I knew it was secure when he failed after many attempts.

Dad determined our farm could support 200 ewes from which we would not only profit from the lambs they would birth but also from the wool we would shear in the spring. Six rams were purchased that would impregnate the 200 ewes. The rams selected were of the Suffolk breed, which a has black head and legs with no wool. We selected this breed for a number of reasons. First, their body structure was broader and more square shaped, which is conducive to more meat. Second, their heads were elongated and not round like some sheep, a shape more conducive for the birthing of lambs. Finally, the wool, though not the best, was still pretty good and would yield a good price. All in all, the Suffolk breed provided a good compromise between having the best of both meat and wool from a single animal.

It was not practical to buy ewes that were purebred Suffolk since they would be very expensive. Instead, Dad spent more money on purebred Suffolk rams with the intent of breeding to obtain the Suffolk characteristics from ewes that were not purebreds. Dad found a few farmers an hour or two away who had about a dozen sheep to sell, but it soon became obvious that since there were not many sheep farms in our mainly dairy and beef cattle agricultural area, the sheep of the quantity Dad needed could not be purchased locally. Dad finally purchased sheep that were birthed and raised out West, which were not purebreds but rather a mixed breed that were hardy and strong from being born and raised on a Western range. We would from then on refer to them as the Western ewes and they composed probably 75% of the flock.

We began purchasing the flock of sheep in the fall. Just like raising the chickens and other farming operations, an annual cycle of repeated specific work began. The first step was allowing the six rams to be with the ewes to start the breeding process. A couple dozen ewes purchased from a local farmer were Hampshire breed and already were pregnant. We kept them separate from the rest of the flock of ewes. They received royal treatment in half of the barn that had straw bedding, good feed hay, and other grain-based food. They would lay around all day, just waiting to deliver their lambs. Dad was concerned, however, fearing it would make them vulnerable to a difficult birthing process. As a result, we would go in to their half of the barn and coax them to walk, which ended up being in a large circle. It got to the point that when we showed up, many would automatically get up and begin their walking exercise.

I can still vividly remember this image of the sheep exercise routine that Christmas season. We purchased a live Christmas tree for a dollar from our next-door neighbor. We cut it down, set it up in the house, and decorated it with our family heirloom decorations. We had a small nativity set with a stable my dad made from a wooden crate with a pitched roof that was painted brown. We placed hand-painted plaster figurines of Jesus in the manger, along with Joseph, Mary, the three wise men, and some plastic cows and sheep. (I still have and faithfully assemble the nativity and use and reuse a small brown bag of hay that originally came from the farm probably 55 years ago. Some figurines still have the twenty-cent price stamped on the bottom and one has Italy stamped on the bottom.)

After completing the Christmas decorating, we went to the barn to exercise the pregnant sheep. Seeing the ewes, hearing the sounds, and smelling the aroma of hay and straw helped me have even a much better picture of that first Christmas scene when Jesus was born in a stable. That was probably my most memorable Christmas on the farm.

Mom and Dad realized they could not continue working a farm with Dad's health concerns, so they were always looking for an opportunity to sell. I wonder if in the back of their minds they were hoping their sons would *not* take up farming because it was such a difficult way to make a living, especially on a small-sized farm like we had. Eventually, the opportunity did present itself when the coal company that mined nearby needed land to build a plant to purify mine water from its iron containing pollutants. Our farm was the ideal spot for this plant because we were where the mine was at its lowest point and so the mine water would naturally flow. In the spring of 1969 when I was sixteen years of age, we sold the farm to the coal company. We also sold the animals and held an auction to sell all the farm equipment and various items that were in the barns and buildings.

My mother, father, brother, and I helped at the auction while the auctioneer sold off each item in a rapid fashion with his rhythmic bidding style. We would help present the items being sold and each of us had a nail bag pouch tied around our waist to hold the money collected and change given to a successful bidder. It was a long day that included not only the auction but also helping to load equipment and purchased items onto trucks and trailers. It marked the end of a life which at the time was the only one I had known. I was just beginning to get more involved with it all, learning the various aspects of the business. It was scary transitioning to a totally new and unknown life and lifestyle.

As farmers, we were never able to go on a vacation since we had animals that needed daily care. Also, the limited income from the farm never permitted us to even consider a vacation. We did have one vacation, however, after my mother insisted that as a family we should have one. We went to Niagara Falls for a two-day holiday and it was truly memorable. Mom's goal was to have a good family life. She had grown up without a father since he died from tuberculosis when she was very young. After that her mother had a hard time taking care of my mother, her two sisters, and one brother. She was passed around between relatives as she grew up. Thus, having a solid family life on the farm fulfilled her goal.

When the farm was sold, we all went on a landmark vacation for our family (Mom, Dad, my brother, and I), a six-week trip out West using a pop-up trailer camper. (It was such a memorable trip that I decided I wanted to have the same experience with my family, which we did in the year 2000.) Life changed completely for me after the sale, though my summer job for the next few years was on a farm where I was paid $6 per days for six days a week if the weather cooperated. My dad got a job at the local hardware store.

My one farm pastime of hunting helped me make good

friends at school and in the neighborhood where we moved since hunting was a big part of everyone's lives. Life changed dramatically with no daily farm chores, but I adapted—though that ethic of farm life and work has stuck with me to this day. My life reminds me of some well-known Bible passages such as Proverbs 22:6 (NIV), "Train a child in the way he should go, and when he is old he will not turn from it." While the hard work ethic remained, what was ahead for me was a life that was far away from what I had known and grown up around. Let me tell you about that journey next.

Chapter Four

Academics

From an early age, I displayed a natural talent for mathematics. As I moved into my junior and senior years of high school, Mrs. Sweeney, the teacher who taught chemistry and physics, piqued my interest by the way he taught these two areas of science—though I still showed much ability and interest in the math subjects. There was an academic track math class with a select 15 students offered on the topic of Analytical Geometry and Intro to Calculus. Mr. Neal, also a very good teacher, taught 15 select students analytical geometry and introduction to calculus, both offered through an advanced track of academic studies.

Everyone in that class was college bound and therefore ready to take the SAT exams, especially the math focused part of the exam. To provide an interesting challenge to everyone in that class, Mr. Neal promised to present a slide rule to the student who had the highest math score on the SAT exam. When I won the slide rule, I realized that maybe I did indeed have a talent in the sciences. After high school graduation, I followed my two brothers' paths and went to a small western Pennsylvanian school named Westminster College where I started as a math major.

I quickly realized that maybe math was not my passion, since taking the first few calculus courses showed me how abstract the area of mathematics could be. At the same time, I was taking chemistry and felt a connection. At the end of my sophomore year, I decided to change majors, but instead of majoring in chemistry, I decided on a combination of physics and chemistry. I thought I could eventually use them to teach at the high school level. I must say the physics department was very accommodating to me and worked out the class schedule for my final two years so I could fit in all the necessary physics and chemistry courses. Two of my four physics professors, Dr Johnson and Dr. Zher, showed a special interest in helping me plot my courses and schedule for those final years. I wondered how many schools went to that level of effort for a single student.

When I changed my major, I had not taken a single physics course, which meant I had to take Introductory Physics I and II and Astronomy in summer school, thus taking a full year of introductory physics including all the labs and problem-solving homework assignment, usually completed in two semesters, in just eight weeks. Needless to say, that was an intense eight weeks of total immersion in physics. In my senior year, I had two thermodynamic courses, one from a chemistry and the other from a physics perspective. They were tough, but while others relied mostly on memorization of equations and formulas, I tried to remember them by understanding how they were derived. That is still how I operate today, for I do not merely memorize a concept presented but first try to understand it from a logical point of view. That is why when I see some of the claims evolutionary science makes concerning the origin of living beings, I saw they did not make sense, raising red flags for me. I will wave many of these red flags in the science chapter.

I completed my double-major degree along with a certification that would allow me to teach in public schools. After teaching one year of high school science and chemistry, I realized teaching at that level, which required some disciplinary actions and measures I was not suited to perform, was not for me. That summer I explored taking some graduate courses to satisfy the extension of my teaching certifications and that search led me to the chemistry department at the University of Pittsburgh.

They informed me that they did not offer such courses but were instead focused on a research-based graduate program. The admissions person for their chemistry department asked me what kinds of things interested me in the field of chemistry. I told him I had been reading about the idea of using hydrogen gas as a fuel in the place of gasoline for cars. His eyes immediately lit up and told me the chairman of the chemistry department, Dr. Wallace, had a large research group and many of them were doing research in metal hydrides that could be used to store hydrogen safely on cars. I was aware scientists were exploring the possibility of metal hydrides being used for hydrogen storage.

Before I knew it, I was accepted into graduate school in the Physical Chemistry Department and signed up to start that fall. This was all accomplished without needing to take any of the required graduate level exams and a few months after graduate students has already been accepted into the program for the fall. Because of my packed class schedule as an undergraduate, I did not have the prerequisite quantum chemistry course, so I immediately enrolled at the University of Pittsburgh and registered for that class which had just begun.

I also had to find a city apartment a few blocks from campus and a suitable roommate to share expenses. All this was quite an abrupt transition for me who was more familiar with rural life and not city living. Looking back, I am amazed to see how all the many different pieces came together for me to even be in graduate school and involved in an upper level of academics that I really was not familiar with because I had no aspirations to obtain a doctorate degree up to that point.

In hindsight from the perspective of my current Christian beliefs I fondly look back and see what I know was God's hand directing and guiding me during those times—even though I did not fully realize it at the time. One of my favorite Bible passages that relates to God's involvement in my life is found in Philippians 1:6: "Being confident of this, that He [God] who began a good work in you will carry it on to completion." I am totally convinced that God orchestrated the details in my life, particularly during the "out of the ordinary" times for me such as moving to the big city, lining up opportunities that were otherwise unlikely because of passed deadlines and the like, to carry out His purpose for me and foster the God-given talents in science I have. To this day I give God the credit for all these and any other accomplishments in my life.

It took me six years to complete the graduate program. The research was multifaceted and therefore took a little longer to complete than usual. My focus involved studies of the thermodynamics of hydrogen reacting with select metal intermetallics, the kinetics or the speed of hydrogen absorption and desorption from these metals, and the examination of the surfaces using Auger spectroscopy. At the time, this spectroscopy was a new surface-sensitive analytical tool that helped me make a correlation between the speed of hydrogen absorption and the chemical make-up of a surface.

Each of these studies required the design and fabrication of scientific equipment. To accomplish this, I learned many skills not necessarily associated with chemistry such as equipment design, acquiring machinist skills of using a lathe and mill machine, designing and constructing electrical circuitry, building high-pressure gas lines and manifolds suitable for hydrogen gas, and even learning the craft of glass blowing. My comfort level of using my hands to design and fabricate the needed scientific equipment came from my years on the farm.

After successfully defending my thesis, I received a doctorate degree in physical chemistry. I remained with Dr. Wallace for a number of years as a research associate during which time I wrote proposals, carried out applied research, and was involved in some scientific publications. Later, the research funds for the metal hydride-related work dried up, after which I taught general and physical chemistry as an assistant professor at St Vincent and Seton Hill., two local small private colleges.

The next year I had the opportunity to be the senior technical consultant for the NASA Technology Transfer Center at the University of Pittsburgh and I maintained this position for ten years. It was fascinating to see the hundreds of technologies developed by NASA and other government labs and their attempts to inform U.S.-based companies of the opportunity to license them for commercialization. After this center closed, my next step was to open a consulting business that filled some of the needs of the closed center using the contacts I had developed over the ten years.

The future included starting and becoming involved with various technology-based businesses that included research and development and manufacturing businesses making highly efficient, permanent magnet-based DC motors, nanotechnologies for inhibition of mold ,bacteria, and algae growth for use in building, aquaculture, and medical materials, along with some research of metalliding for modifying metal surfaces. My career path has entailed a wide variety of subjects and topics for which I had to learn and adapt, a skill that proved useful in my study of the origins of living beings. These were years not only of significant intellectual growth but also spiritual growth, which I will chronicle for you in the next chapter.

Chapter Five

Spiritual Growth

Mother and Father were totally committed to see their children faithfully attend church with them every Sunday at Rehoboth Presbyterian Church, a small country congregation founded by a circuit rider preacher in 1778. It was a church my grandfather and grandmother, also farmers by trade, faithfully attended. My dad served as an elder of the church and I can still picture him preparing on Saturday evening to teach a Sunday School class the next morning. To be truthful, spending time in the Sunday School classes and then the worship service was not very interesting for me. I do remember some of the old hymns we sang, however, and they still touch my heart.

The big moment in my childhood came in the summer between the third and fourth grade. Our church emphasized the importance for the children to attend a one-week church camp, even providing funds to help defray the cost. That week made an impact on me that I did not fully realize until later for that is the first occasion I had made a decision to accept Jesus as my Savior. With all the Bible lessons, songs sung, and nightly chapels throughout the week, I actually sensed God and responded. That made quite an impact as I got older, I wholeheartedly committed to volunteer at that same camp as a counselor and as I had experienced, I presented lessons and talks from the Bible to the campers. This is something I did for ten years during my college and graduate school summer break.

When I was 14, I was then eligible to become a church member after attending a communicant class. I recall the pastor coming to personally meet with me at our home. At the time, I knew church membership was a big deal and I made it a bigger deal than what others fully realized. When asked by the pastor if I wanted to join the church I knew, my response was something like, "I think so." Mother was taken aback by my answer. I was only a kid and one reason for saying this was not due to apathy, but rather because it truly was a weighty decision that I did not want to take lightly.

When going off to college, I still intended to go to church, but to a church I felt had a bit more to offer. I attended the college-affiliated church my first two years but then branched out to a church in town whose preacher truly spoke to me from the Bible about who God was, which started the acceleration process of my spiritual growth. I actually began reading the Bible and praying to God more sincerely. My first job was teaching in a rural high school general science and chemistry classes. At the time I was a pretty quiet individual, and for me to choose the teaching profession where I would need to speak in front of classes everyday was an amazing step. I started to depend on God more often and would actually silently pray before class asking for His help and from this, I saw my first answers to prayer. Living alone as a single person, I had free time at night, which I used to read through the entire Bible.

As I explained above, my high school teaching career only lasted one year, since the next year was when I entered graduate school. Little did I realize that aside from obtaining a doctorate degree in physical chemistry, the bigger impact on my life was in my spiritual walk. I grew tremendously during those years. Bellefield Church is where I attended, became a member, then a deacon, and finally an elder. It is also where I met the love of my life, Janet, who became my wife. It is curious that the first Sunday I was in the city, I planned to go to a Baptist church, thinking that would be something different to try.

As would have it though, that morning I was uncharacteristically early walking to church. Since I had extra time, I thought maybe I would go to Bellefield Presbyterian Church, the denomination with which I was most familiar. Little did I know what a huge life-changing decision I had made, and I know God's hand was on me that morning. Bellefield had an active outreach to the students at Pitt through the Coalition for Christian Outreach. I began to attend small group Bible studies, not only those sponsored by Coalition, but also by the Pittsburgh Regional International Students Ministry. The latter is where I began to develop many friendships with people from other countries, learning much about their cultures in the process.

Eventually I took on a student leadership role for the Coalition ministry where I would co-lead small groups that would have a Bible study and also perform some service in the community. After the first couple of years of co-leading with different people, I partnered with a man named Bob to lead a group that would stay together for the next four years. Bob and I became friends and brothers in the faith, meeting once a week at 7 a.m. for breakfast and prayer. The group became very close as we conducted in-depth studies of the Bible, prayed for each other, and shared each other's burdens. It was how the church should actually function.

I became a member of the church. The leadership at the church must have taken notice of my spiritual growth and leadership and nominated me to serve as deacon, a group dedicated to service and helping the poor. Since the church was in the city, there was no shortage of people needing help, and the role cultivated and developed my talent for helping people and strangers. We even set up an Open Home Ministry for those who arrived at the nearby Presbyterian Hospital to receive a transplant of some sort. At this point in their medical care, the family there for support could not afford to stay in a hotel. Therefore, the members in the church who had a spare room to share provided the temporary housing needs for these families.

We saw God work through this ministry as individuals with needs being met by this medical community were also being loved and touched by God. This was before the Ronald McDonald House program was thought of or started. A few years later a Ronald McDonald house did open in the community that negated the need for the Open Home Ministry. After my term as a deacon came to an end, I was nominated and elected as an elder, probably one of the youngest to serve in that office.

Bellefield Church played a significant role in my tremendous spiritual growth during my graduate career. However, there were two other events that helped spark my spiritual growth. I became good friends with a man named David whose family attended the church and who had tremendous knowledge of geology. David often talked about the science of geology naturally intertwining with the Bible. He networked with a number of other scientists, medical professionals, and engineers who thought the same way. They started a Pittsburgh Creationist group and David invited me to attend their monthly meetings. There would be a few dozen professionals (medical doctors, engineers, and geologists) who would attend each month and someone in the group would present a talk related to their science profession, linking their findings to the Bible. For example, the geologist would connect the different sedimentary layers to the worldwide Noahic flood described in the Bible. Seeing the powerful, solid scientific evidence they presented had me become more open-minded and very interested in their knowledge and way of thinking.

Another significant happening that shaped my thinking as a Christian was around an informal Bible study that naturally came together in the Chemistry Building at the end of the eighth-floor hallway, attended by no more than 10 graduate students. At this time, a curious post-doctoral researcher and brilliant scientist named Satoshi from Japan had joined our research group. As generally occurred in university research groups across the country, the first job or position of a newly degreed doctoral students would be a researcher in a group that fit their background and interest.

Satoshi's interest in magnetic materials matched part of Dr. Wallace's research group focus in studying magnet materials. Eventually Satoshi was invited to the eighth floor Bible study. Although he did not actually participate or add to the discussion, he faithfully attended over a period of a few months. Some from the group suggested maybe someone should talk with him one-on-one to see if he wanted to know more and possibly become a Christian.

Since I was in the same research group and had actually began developing a friendship with him, I volunteered to go and have that discussion. Much to my surprise, I found out that sometime since he had begun attending the Bible study, he had made the commitment to become a Christian. Needless to say, I was excited and welcomed him to the faith.

Later I wondered about his reasons for becoming a Christian. A few days later I went to him and asked why he had made that decision. His answer amazed me and set me on a slightly different path in regard to growing my own faith. Instead of providing an emotion-based answer of how he was moved to convert, he simply said he did so because it was so logical—and this coming from a man whose country's population was made up of only 1% Christians, the vast majority of Japanese holding to a Buddhist faith.

What I came to realize is that if what is said in the Bible is actually from God (and of course it is), then there should be evidence to back it up. In the end, faith is still required, but there is enough evidence available that blind faith is not necessary. At the beginning of my presentation of the *God Discovery Banquet*, I use the Grand Canyon as a metaphor to explain this concept. Let me explain.

The first slide presented at the Banquet is that of a familiar view of the Grand Canyon that shows the awe-inspiring canyon with its multiple-mile gap from one side to the other. I explain that for many people, the Canyon represents their view of belief or should I say their inability to believe in God. Their leap in faith would be the equivalent to the size of the canyon. I change the slides showing different pictures of the Canyon as one travels upstream and those present can see that the canyon gets narrower. I liken this to what happens when one honestly investigates the possibility of God and the potential evidence that may exist, the leap becomes more reasonable. Then I bring them to a point upstream where the leap across a small stream is doable.

There are many who have taken the challenge to investigate the evidence and it eventually resulted in their coming to faith. One in particular was Josh McDowell. He actually started out to disprove Christianity but what occurred was the exact opposite. The evidence he gathered was so compelling that he put together a series of books that presented the evidence he uncovered. *More than a Carpenter* is a book that provides a summary of his findings. This was one of the first books I read that helped me solidify my faith.

His other series of books titled *Evidence that Demands a Verdict* provides much more in-depth evidence for those who want to dig deeper into the subject. A more contemporary author who had a similar experience is Lee Stroble. Lee has written numerous books such as *The Case for Christ*, *The Case for Faith*, and *The Case for Creation*, to name a few, that also provide much evidence for the validity of the Christian faith. Let's now proceed so I can give you more of what I present in the Banquet concerning science and its indication that God does indeed exist.

SECTION TWO

God and Science

Chapter Six

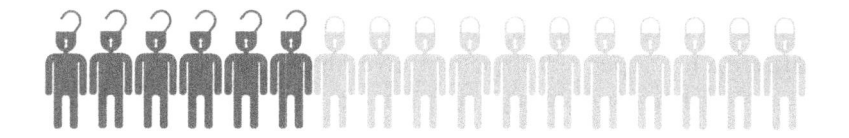

Science Points to God the Creator

The following explanation will show how impossible it is that life came into existence by accident or simply by chance. Let me start at the very beginning of the evolutionary process with the formation of the most simple and essential biochemicals. From there, I will show how equally unlikely the next step was, which is the formation of proteins which led to the assembly of the cells of a body. From there, we will look at the DNA super-molecule where great intelligence would be required to synthesize DNA. Finally, we will examine the unbelievable complexity of the physiology of the human body, which once again points to the need of much greater intelligence in its construction.

I am convinced that all we see and experience in nature comes only through the creation by an intelligent being, namely God. Conversely, many people who look through the eyes of science try to explain where we came from or how everything we see and experience came into being through the process of evolution. The aspect of evolution that is generally looked at is what I would call macro-evolution—the evolution of one species to another. The most common link searched for is that between man and the ape.

The connections made through fossils, bones, and other life artifacts are incomplete and composed of scattered items of study from the distant past. Proponents of evolution use these items to piece together and put forth a narrative to support their theory. Examples of things used to prove the human/ape link are the shape of the head or fossilized bones from various discoveries. There is never a compete structure or skeleton found. The following are a few aspects of these narratives that concern me.

First, people from various backgrounds and cultures today have a wide variety of physical and bodily forms which appear to match some of these excavated ancient remains. Also, almost all the remains of prehistoric finds are fossilized or bone fragments, meaning the majority of what makes up a human, animal, or plant decayed and no longer exist. Thus, the missing parts are created through one's own thoughts mixed with imagination and creative thinking, the end result of which more likely does not accurately represent the actual being that existed. We have no actual photograph of a prehistoric human being, just drawings from someone's own perception or idea of what they could have looked like.

Then there is the logic deployed that if there is a commonality of physiology between species involving the existence of a backbone, nervous system, or circulatory system, with another simpler or smaller species, it is proof that simpler or supposedly more primitive ones were the precursors to a higher life form. Simply stated, the postulate that the simple living species evolved into a more complex species that possessed and maintained the same fundamental structures.

The opposite of this thinking would be that a master designer, someone like God, developed a well-engineered structure for a specific use or task, then utilized this same structure to develop other species that share the same underlying structure. A case in point would be items designed by engineers over the years used in transportation or the movement of materiel and people. They all have a frame of some sort that generally contains the same type of structural members. There are a variety of wheels and tires found in all vehicles, which can even be found in tracked vehicles such as bulldozers, skid loaders, military tanks. All those incorporate wheels to move the vehicle itself powered by some type of engine to produce a circular motion needed to move the vehicle. However, I digress from where I will be focusing my attention.

I am not a biologist or a paleontologist. My training and education have been in the areas of more fundamental sciences—physical chemistry and more specifically the study of thermodynamics. There is a hierarchy of knowledge in the sciences, starting with mathematics, then physics, next chemistry, followed by biology and finally on to other sciences like psychology and sociology. This does not suggest that one science is superior to another in terms of importance or its impact on a society. After all, many of the advancements in biology and medicine have had a tremendous effect in the world, resulting in great advancement in health, a better quality of life, and longer life spans. Discoveries in microbiology of germs, bacteria, and viruses along with an understanding of how they live, procreate, and can then be controlled have greatly influenced our lives today.

However, the law of physics must follow the rules of mathematics. Two plus three always equals five. Nothing in physics can dictate that two plus three equals four in the world of mathematics. The same goes for the connection between physics and chemistry as well as between chemistry and biology, for the laws proven true in chemistry cannot be disregarded when studying and understanding biology. I fear we have lost this understanding of the connectivity between these sciences in some cases.

For example, the laws of thermodynamics in physics and chemistry do not appear to be clearly understood in biology's subfield of evolution. Those intimately involved in these studies seem to be too far removed from the science of thermodynamics. Let me explain why I make that statement. Thermodynamics can be understood by breaking down the word itself into two parts: *thermo* and *dynamics*. *Thermo* refers to heat or energy and *dynamics* involves the changes that occur due to interactions within a system. Therefore, thermodynamics involves the study of the change of the heat or energy of a system due to any interactions that occur in the system. Striking a match, an interaction, within a room yields the formation of light and heat from the stored chemical energy (the heat or energy) that existed within the match head and the matchstick.

Most of the time, I will attempt to present the discussion

of science and more specifically thermodynamics in layman's terms, though I will delve deeper into the science in a few instances. In those times when I dig deeper, if it is hard for you to comprehend, I suggest you just pass through and not take the time to try and understand the details. If it's too technical, seek to have an overall sense of the concept being presented. At the end, I will summarize the point I am making.

There are three basic laws of thermodynamics. All that exists in the universe, with a focus on the interaction of energy and matter, are found to behave within the confines of these laws. Only the first two have relevance to what follows.

THE FIRST LAW OF THERMODYNAMICS

The first law states that heat and/or energy cannot be created or destroyed but can only be changed or converted to other forms. When considering the example of lighting a match, we may reflect at first glance that the energy stored in the match head and stick was being used up or destroyed upon lighting the match. That way of thinking is contrary to the first law. Instead of losing or using up the energy in the matchstick, it was converted to other forms of energy, namely heat and light that were not lost or destroyed, but rather dispersed in a diluted form to its surroundings.

The heat and energy that were stored as chemical energy in a concentrated form in the matchstick still exist seconds, minutes, hours, years, and even into eternity, but they are in a more randomized, dispersed form. The chemically stored energy of the matchstick is converted to the other forms of energy consisting of light and heat or vibrating atoms and molecules. Light can strike a wall and will ever so slightly increase the temperature of the wall. In other words, the total energy content of the universe as defined by the First Law does not and never has been used up or destroyed. It has and is just changed from one form to another—chemical to heat and light, light to electrical, heat to chemical, etc. If anything, energy as a whole is not as concentrated over time but becomes more dispersed and diluted, and therefore more unusable. This diluting of energy is actually part of the Second Law of Thermodynamics

The Second Law of Thermodynamics

The Second Law is a little easier to understand and grasp in terms of our daily existence. I will use this Law almost exclusively as I put forth a case for there being a God who has made all living things. Simply stated, when looking at the dynamics of change within a system, everything that spontaneously changes will tend to move to a more disordered state. To obtain a more ordered state what I term as *ordered energy* needs to be inserted into the system. The thermodynamic term that is used to identify and quantify the disorder of a change within a system is called *entropy*.

When a system experiences a change from an ordered state to a more disordered state, the system is said to have an increase of entropy. The entropy and the heat involved in a system like the one found in the lighting of a match have actual quantifiable numbers calculated to determine the amount of heat and entropy change that occurred. Before the match was lit, all the energy was in a more ordered state, concentrated in a chemical form on the head of a match. Upon lighting the match, the energy is dispersed from the concentrated match tip outward to its surroundings in various energy forms, which is a more disordered state.

A few common examples will help to illustrate the Second Law and provide you a much greater understanding. Figure 2 shows three examples where an increase of entropy occurs. The first is a photo of a desk, a desk that could very well be my desk. It is messy and from all accounts is a picture of disorder where entropy has increased over time on the desk. All work desks will trend toward messy and unorganized if left unattended throughout a workday. That's why many people will make time at the end of the day to devote intelligent human energy to order, organize, stack, shelf, or file the papers and items on their desk.

The second picture relates to the financial markets. The same rules apply in that world since if a financial analyst does not pay attention to plan and direct how funds are invested, they will likely not yield good results. The last example in the figure can be related to materials and chemistry. The picture is of salt and pepper placed in two separate Erlenmeyer flasks. They are in sort

Examples of Entropy at Work

- House clutter
- Economics and finances
- Mixing together salt and pepper

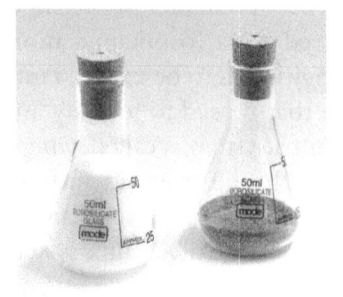

Figure 2. Common everyday examples of entropy

of an ordered state—salt in one flask and pepper in another. If one would take and mix the two together, it would represent a more disordered state having increased entropy.

If one wanted to separate the two from this mixture and put them back into a more ordered state with less entropy, an organized process would need to take place requiring ordered energy. It would be a tedious approach to take tweezers and physically remove each grain of salt and put it in one pile while doing the same for each pepper flake—a labor-intensive but orderly approach.

A more sophisticated technique would be to take the mixture and add it to water. The salt grains would dissolve while the pepper grains would remain intact since they are not soluble in water. Next, we could take filter paper, place it over a funnel, and pour the mixture into it. The salt solution would pass through the filter while the pepper grains would be caught in the filter paper. Once the filter paper is dried, the pepper grains could be retrieved. The salt solution could then be placed in a dish from which the water evaporates over time. We could even apply heat used to expedite the evaporation, leaving behind salt crystals. These crystals would then need to be scraped off and ground into salt grains. This chemical approach, though more elaborate, still requires a tremendous amount of organization and ordered energy to accomplish this task.

The above illustrates that to obtain an ordered state or a state with less entropy, a great deal of energy needs to be expended. It cannot be just any energy, however, but must be ordered energy. If this ordered energy is not expended, then the Second Law prevails so that processes and events naturally go to a disordered state. My experiences growing up on a farm, specifically with all the specific work and chores done, showed me there needed to be a great amount of organized energy put forth to bring about the desired results of raising cattle, chickens, sheep, and lambs, along with growing corn, grain, and hay. This included complex farm equipment; daily, weekly and annual routines; and human energies invested in planning and coordination—all to counter entropy.

If this work was not done, the farm as an organized and productive entity would collapse and everything would fall into disarray. It is remarkable how little time it takes for fields to go back to a wild state and buildings fall into disrepair and even collapse when a farm is abandoned. You probably have numerous examples of when you have had to work to counteract entropy in your life or job, or perhaps a time when you did not and the results were predictable—things went into disorder and not order. That is why I see the concept of entropy (the Second Law of Thermodynamics) is probably the most easily understood of the three laws of thermodynamics. We all encounter it, easily recognize it, and struggle to overcome it on a daily basis.

Chapter Seven

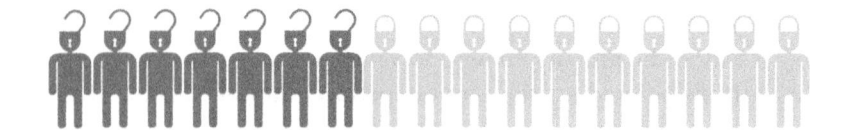

Evolution – An Implausible Theory

It is a fascinating exercise to overlap two graph plots that shows what one should see when comparing the overall system order of all living species combined from the point of view of the Second Law of Thermodynamics (entropy) and then from the view of the theory of evolution. Figure 3 show these generalized graph plots. The theory of evolution starts with random atoms and transitions to simple molecules, then to more complex, organic-based compounds such as amino acids, which are formed through trial and error over a long period of time. Over extended periods of time these join together to make more complex organic polymers such as proteins, fatty acids, and other biochemicals necessary for life. Those then join together to form extremely complex molecular systems such as RNA and DNA, followed by cell creation of a vast number of different types (skin, muscle, nerve, blood, bone, cells). These then form infinitesimally complex organisms that have an unbelievable interconnectivity of the various functioning cells which yield a living species with a myriad of diverse biological and physiological functions, all working together to promote the existence and propagation of the species.

Comparing System Order Trends Expected for Chemical Evolution and 2nd Law of Thermodynamics

— — 2nd Law Thermodynamics Trend

- - - - Chemical Evolution Theory Trend

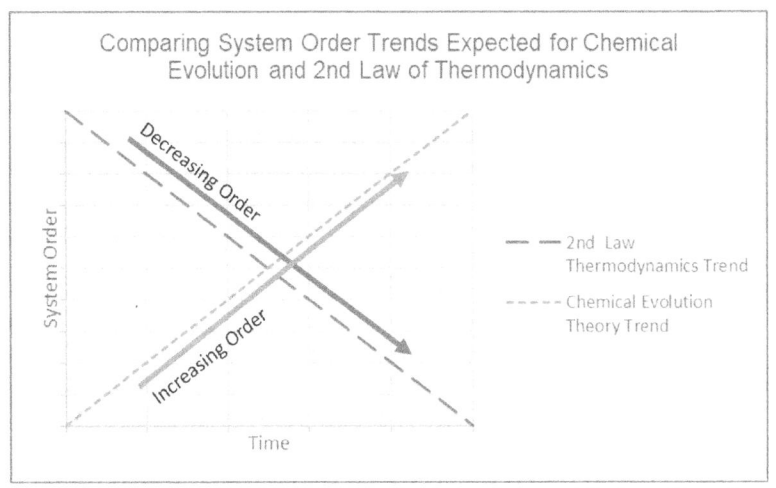

Figure 3. Plot of trends one should see for chemical evolution and the Second Law of Thermodynamics (author's graph)

The graph or plot in this figure represents the evolutionary progression initiating the progression of species formation starting with time and the amount of system order at zero. The time is presented as the horizontal axis of the graph and the system order of all species combined presented as the vertical axis. As time progresses from zero, where only simple unorganized atoms exist, to later times, the order of the overall system simultaneously increases upward in the vertical direction, resulting in a line drawn from the starting point at the lower left-hand corner (i.e. the point where only simple unorganized atoms exist) towards the upper right-hand corner (i.e., where all living species exist). The upper right-hand corner is where the overall combined order has increased dramatically from its starting point at time zero.

The other plot or graph shown in this same figure is how one would explain what would be expected if the Second Law of Thermodynamics is applied. As time goes on, the overall system order of all species combined would start out at time zero at a maximum and with progression of time would be expected instead of increasing to be decreasing and the entropy would increase, resulting in more disordered systems. As you can see in the graph, this plot starts time zero at the top of the vertical axis with the maximum order and drops as time increases to the right.

As you can clearly see, these two lines are moving in the exact opposite direction. Thus, evolution and the Second Law clearly contradict one another.

EXAMPLES OF ENTROPY

Each of us are actually living examples of this reality. As we grow older, our bodies deteriorate and wear out, needing replacement parts such as hips, knees, and hearts until the body fails to the point of death. At that point, entropy continues as the body decays rapidly, with all the tissue and complex molecules breaking down to simpler and less complex molecules. These results of molecules that are the product of decay are in a more stable energy state that is more disordered. This disorder is realized in the more randomized smaller sized product molecules. The disorder is also found in the subsequent energy state. The First Law states that energy is not created or destroyed but rather transformed. Before decay, the complex molecules actually have a lot of energy within the bonds of the molecules, which is released and results in more stable, simpler molecules that are scattered and with less available energy.

A good example would be the burning of coal. Coal consists of long chains of organic-based hydrocarbons. When burned, chemical energy contained in the bonds between atoms making up the complex hydrocarbon molecule is released in the form of heat and light energy yielding much simpler molecules of mostly water (H_2O) and carbon dioxide (CO_2). The higher available chemical energy stored in the bonds of the coal molecules was not destroyed but transformed into lower available chemical energy in the bonds of hydrogen to oxygen in water and carbon to oxygen in carbon dioxide atoms—plus the heat and light that were radiated from the process.

The overall final energy state is scattered and not localized, ending up in a more disordered energy state divided between bond energies of H_2O and CO_2 and the heat and light being dispersed outward and away. This randomized final state is the exact opposite of the starting state which was one that is more ordered in a more compact space of the initial complex and organized long-chain molecules of coal.

At this point one may ask, and rightly so, why we see ordered systems continually coming into existence around us. Chicken eggs will yield complex baby chicks, a grain of corn, or barley will sprout plants that yield many more seeds, and husbands and wives will raise babies to become more complex toddlers, adolescents, and eventually adults. To answer this question, we know all these living things do not last. Entropy catches up with them all eventually. Plus, over the course of long periods of time, defects will and can be incorporated into the genome, which is the main reason the marriage of close relatives is not advised. The likelihood of a defective genome due to the effects of entropy will actually be realized and an actual defect in the child of closely related parents is greatly increased with parents that may have more similar genomes. This occurs when attempting to breed pure-bred dogs, resulting in effects such as hip dysplasia, eye issues, and shortened lives.

Therefore, what we have is a dichotomy. On one hand, there is order in living things passed from life to life, but during the course of life there is a constant battle as entropy tears down a living system to the point of death, an inevitability for all living things. What is actually going on in terms of thermodynamics? Let me introduce a thermodynamic equation which in this case is known as a state function.

A state function or equation calculates the thermodynamic values and quantities gained or lost over the course of a change or reaction in a system. The key characteristic of a state function is that these calculated values and quantities do not depend on the path or route required to go from the starting materials and parameters to the end products of interest. If there are a variety of paths, ways, or intermediate reactions to get to the final product, the end thermodynamic values will be the same. No matter which path is taken, entropy is one of the thermodynamic values derived from a state function.

A simple illustration in regard to a state function regarding entropy with two different ways or paths is the example I provided above—separating a mixture of salt and pepper. If you remember, I presented two possible ways of separating the mixture of salt and pepper particles. The mixture of salt and pepper represented a state having more entropy when compared to the salt and pepper that were separate in two different containers. The first path for separating the mixed salt and pepper into the ordered, lower entropy state was with tweezers. The second path was dissolving the salt in water, filtering out the pepper and then evaporating the water to obtain the salt. The entropy change from the salt and pepper mixture to the separated state of salt and pepper would be exactly the same, regardless of the method deployed to separate the salt and pepper.

A more complex illustration would be the production of amino acids. The basic and simplest biomolecules needed by living things are the 20 different amino acids which are the building blocks in the construction of proteins. The amino acid building blocks are assembled together to make longer protein molecules that are used as the building material for living cells. One possible path for making amino acids is for a living organism to fabricate them with the various biochemical reaction steps used within a living cell. That is the natural way of producing amino acids. Another path would be to artificially fabricate the amino acids within a chemistry lab using beakers, flasks, separation funnels, and solvents to end up with the same amino acid. In both cases, the change in the thermodynamic values of energy and entropy from the beginning reactants to the final amino acid product would be exactly the same, whether produced naturally within the cell or artificially in a laboratory.

Please allow me to present one more example of the production of a chemical substance both naturally and artificially where the entropy change is the same. During my time as a high school teacher, I had my students perform an interesting organic chemistry lab exercise of synthesizing an organic chemical called esters. That is the chemical one can smell and taste in various fruits and spices. We would synthesize various esters by mixing specific carboxylic acids with selected alcohols.

The end results were products that smelled nothing like the starting carboxylic acid and alcohol reactants but rather like various fruits and spices such as bananas and spearmint. Again, the entropy change to produce these fragrances of fruit and spice is the same, whether produced naturally by a tree and plant or produced in a high school lab. By the way, when a food product states no artificial flavoring, it means the flavor-based compounds were extracted from a fruit or vegetable and not synthetically produced like we did in the chemistry laboratory.

Time for Some Math

The next thermodynamic state function I will present contains one of the variables in the function, entropy. This is the only mathematical equation I use in this book, so for those who dread math and science, don't be alarmed. I will keep it simple and short. If you have difficulty understanding my explanation, please just plow through the next few paragraphs to the end where I will explain the key end point to this equation. This equation shows how the unavoidable increase in disorder or entropy occurring all around us is counteracted to obtain the order necessary for life itself. The function is termed the Gibbs Free Energy. The presentation and details of the Gibbs Free Energy (taught in any college chemistry class) for our purposes will be kept at a simple level of understanding the interaction of entropy and the energy of a system. The simple and non-calculus form of this equation is:

$$\Delta G = \Delta H - T\Delta S$$

Δ represents the difference in the value from the starting value of the initial reactants to the final value of the end product of interest. The = represents the change of G that equals the values calculated after the equal sign. And − is simply a minus sign. G is the Gibbs Free Energy value (from any chemistry class) and therefore ΔG is the difference of Gibbs Free Energy calculated for the change of the system. The H is the enthalpy of the system which by definition is the thermodynamic quantity equivalent to the total heat or energy content of the system. T is the temperature of the system and S is the letter used for entropy with ΔS being the entropy change of the process of interest. So, what is the big deal about G or more importantly ΔG?

In order for a system or chemical reaction to be thermodynamically favorable or for the reaction to occur, the value of

ΔG needs to be a negative number. If you are still with me good, if not, don't worry, just press on. We are almost at the end where I will explain the overarching concept I am putting forth. Let us say we are looking at the process for the formation of a more organized, complex protein chain from simple starting random reactants of nitrogen, hydrogen, and carbon-containing molecules. Moving from a disordered state (simple nitrogen, hydrogen and carbon molecules) to a more ordered state (proteins), meaning the entropy is decreasing, makes TΔS a negative number. (Side note: T uses the Kelvin scale of temperature, not Centigrade or Fahrenheit. There are no zero or negative Kelvin temperatures.)

Remember, entropy is a measure of disorder, and a positive number indicates disorder is happening and a negative number shows the opposite is happening— order is occurring. Combining this negative TΔS number (making a protein molecule, a more ordered state, makes this negative) with the minus sign in front of the TΔS in this equation actually makes this a positive term or number for this part of the equation. To have this reaction or system proceed according to Gibbs Free Energy, the ΔH for this equation has to be more negative to counteract the positive part of the equation, meaning that heat or energy content has to be given up or provided by the system.

At first glance that would not appear to be the case. The amino acids have a greater heat content than the simple starting molecules. What is happening? Heat or energy content is injected somewhere along the multi-chemical reaction process that produces the amino acids. The living cells within the body interject the needed heat or energy content through their production of the amino acids. They act like an amino acid factory. Where do they obtain the necessary energy for the heat or energy content required? They do so through nutrients received in a complex process from food.

Ultimately, the energy obtained to raise this energy content in the food by the formation of complex energy containing organized food molecules come from the sun. It's all grounded in the important use of chlorophyll in photosynthesis to grow plants. Therefore, if the sun is included within the whole encompassing system, the overall ΔG is negative, taking into account the changes on the sun are a loss or giving up of ΔH (making a negative number) along with a greater increase in entropy (a positive ΔS giving a negative $T\Delta S$ when taking in account the minus in the equation).

If you made it through the above reasoning, congratulations. If not, no worries. The end conclusion is that energy injection, which originally can be traced back to the sun, is needed to counteract the increase in entropy and disorder that would naturally and inevitably occur. The bottom line is that energy is needed to create life. It is curious that the energy from the sun for making the needed reactions possible to create life is not the sole criteria for sustaining and growing life.

If all the necessary elements were put together in a container, no one would believe that just by shining sunlight on these most basic elements, such as graphite carbon, gaseous hydrogen, oxygen gas, and nitrogen gas, that a living organism would suddenly come into existence. Anyone with no background or knowledge of thermodynamics or any of the sciences can understand and accept this. This concept and way of thinking have been demonstrated in a well-known if bit dated Hollywood movie, *Frankenstein*.

The goal and claim of Doctor Frankenstein in any of the Frankenstein movies was that he created life. But to do so he used body parts from deceased bodies, sewing them together and providing energy from a lightning strike to restart the life. The starting part in his creation was the already existing complex and perfectly constructed pieces for the assembly of a physiologically complete person. He did not use carbon, water, nitrogen, and hydrogen. I know that this is merely a science fiction movie, but the point is the doctor used a means of creating a monster from body parts that in people's minds is more plausible due to our basic understanding of how bioprocesses work in the real world.

What does happen in the propagation of life is that sunlight is received by plants and through the use of chlorophyll and the process of photosynthesis the plant converts solar energy into chemical energy. This is in turn stored in plants and utilized in plant growth, requiring simple carbon dioxide gas, water, and nitrogen-containing molecules to yield plant growth. Taking these simple molecules, plants convert them into much more ordered (lower entropy) with more energy containing biomolecules, utilizing the energy provided by the sun to do so. The plant is actually behaving like a machine that takes in solar energy along with simple molecules such as water and carbon dioxide and makes the more complex biomolecules.

Thus, the necessary requirement as dictated by the Gibbs thermodynamics function of needing energy to produce ordered systems is satisfied. To accomplish this, however, the required energy is manipulated and transformed through what I label a machine, in this case a bio-machine. These higher energy and lower entropy biomolecules of the plant or the fruit of the plant (created by the biomachines, i.e., the plants) is what is needed as food or nutrients by living animals to produce the lower entropy and higher energy containing amino acids, proteins, and other biomolecules necessary for the physical growth and day-to-day functioning of that animal.

Chapter Eight

Challenging Chemical Evolution

The area of science that the reasoning above actually challenges is the field of chemical evolution. Chemical evolution is the absolute first and necessary step that must occur before one can even talk about any of the higher levels of evolution such as the evolution of land animals from fish, the evolution of birds from reptiles, and the evolution of man from an ape. If one cannot show or demonstrate that the 20 relatively simple amino acid molecules, the essential building blocks for life, are reasonably and spontaneously made from the simple elements of carbon, hydrogen, nitrogen, and oxygen, then there is a major flaw in regards to the viability of the evolutionary theory.

Before we proceed any further, let's examine the actual

$$\begin{array}{c} NH_2 \\ | \\ R-CH-COOH \end{array}$$ **General Structure of Amino Acid Molecule**

makeup and structure of amino acid molecules. The 20 essential amino acids molecules for making proteins have two things in common. They all have the two functional groups of an amino group and a carboxylic acid group. The figure on the previous page is a general representation of the molecular formula of amino acids.

The NH_2 in this formula (N is a nitrogen atom and H_2 are two hydrogen atoms that are bonded to the nitrogen atom to make the amino group) is the amino group and the COOH is the carboxylic acid group (C is a carbon atom, O and O are two oxygen atoms, and H is one hydrogen atom, all bonded together to make the acid group), both linked together with the methyl group CH. The R can be a variety of carbon (C) and hydrogen (H) based groups composed of linear chains of carbon and hydrogen or ring groups of carbon and hydrogen atoms. A specific arrangement of a carbon and hydrogen atoms linked together for the R part of this amino acid formula is what determines what the amino acid is. The simplest substitution for R would be hydrogen (or H) yielding the following molecular formula that is the smallest, simplest, and most elemental amino acid called glycine.

$$NH_2$$
$$|$$
$$H-CH-COOH \quad \text{Glycine (Simplest of 20 amino acids)}$$

There have been experiments to show that the injection of energy into a closed system containing the necessary starting elements could yield amino acid molecules. One of the first experiments that would show the plausibility of amino acids being created from the most basic elements was performed and reported in 1953 by researchers at the University of Chicago (Thaxton, Bradley & Olson, 1984). This book provides extensive detail and in-depth explanation of the results and thermodynamics involved.

Stanley Miller, a graduate student working under Harold Urey, a Nobel Laureate, set up an experiment that was meant to simulate an early earth atmosphere and environment using the gases (along with their chemical formula) of ammonia (NH_3), water vapor (H_2O), methane (CH_4), carbon monoxide (CO), and hydrogen (H_2). A diagram showing the glassware apparatus that was built to carry out this simulation and attempted synthesis of amino acids is provided in Figure 4. To create water vapor, a flask containing water was incorporated into the apparatus and heat was applied to the flask to generate the vapor (Thaxton, p. 22).

The other elements listed above, which are naturally gases at room temperature and pressure, were also injected into the apparatus. An important item to notice in this experiment is the omission of oxygen and nitrogen gases that together compose 99% of our current atmosphere—an important point we will discuss later (Thaxton, p. 76). A trap part of this device, shown at the bottom, is used to collect any product that may be produced. Many types of energy such as electric sparks (to mimic lightning), ultraviolet radiation, heat, and sound were introduced into this system (Thaxton, pp. 24-33).

So, what were the results? There was a great deal of enthusiastic excitement as the graduate student Miller presented his

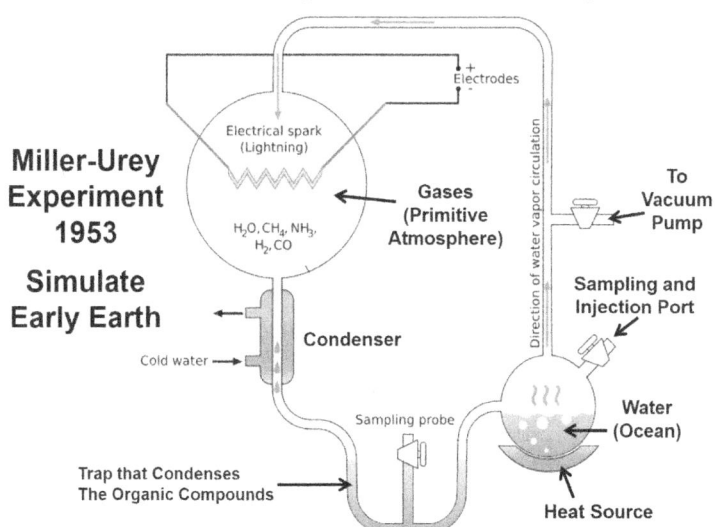

Figure 4. Drawing of Miller-Urey experimental set up

results to Urey because they were able to create some amino acid molecules. Through the course of their experiments, they synthesized only 10 (Thraxton, Bradley & Olson, 1984, p. 24) of the 20 amino acids which are building blocks for protein molecules needed for life to exist, all collected in the trap containing water in extremely small concentrations. In addition to these 10 amino acids, around 26 other amino acids that are not used in living structures were found to be made (Thaxton, p. 53), which would greatly hinder the successful formation of a protein chain of the allowed amino acids by substituting any of the unallowed 26 amino acid molecules. Regardless, work continued with the hope that the remaining 11 amino acids could be created in larger amounts once the parameters and conditions, such as temperature of the water bath, variations of intensity, and types of energy injected, could be fine-tuned to the optimum settings.

Though other experiments yielded other amino acids in very diluted amounts, not many other encouraging results have been reported. Yet the science community in general appears to accept these results as proof that chemical evolution is possible and therefore is the initial step necessary to obtain life according to the theory of evolution. Not many will question this even though *The Mystery of Life's Origin* points out many problems and issues that exist with this experiment and the results. These cast a cloud of doubt over the likelihood of chemical evolution spontaneously proceeding based on these experiments. I will present in layman's terms the problems and why they are really the death knell for the concept of chemical evolution.

THE PROBLEMS

The first obvious issue is that not all of the necessary amino acids were synthesized during these early earth-type experiments. For those amino acids that were made, they required stringent, very limited conditions. *The Mystery of Life's Origins*, published in 1984 or 30 years after Miller attempted the first earth simulation synthesis, provides a table (Thaxston et al, 1984, p. 36) of nine different research groups that performed the early earth simulations synthesis experiments using different parameters where 16 of the amino acids of the needed 20 were found to have been made. Without the other four amino acids, living organisms could not have evolved.

The most successful research group of the nine listed in this table, Matthews & Mosser, made 12 amino acids of the 20 needed to build protein molecules. Two of the 12 amino acid molecules were found not to have been made by any other of the eight research groups. What experimental condition did they use to differentiate them from the other eight research groups? Their experiment used as a medium anhydrous liquid ammonia, which when translated into laymen's terms is pure ammonia for which all traces of water have been removed. All the other eight groups utilized aqueous or water containing solutions, which would appear to be a more realistic simulation of early Earth conditions considering that more than 70% of the Earth's surface is water.

The next significant problem was the low numbers of amino acid molecules that were made. This is because the reactants were more desirable in terms of being more energetically stable and at the same time having higher entropies. The equilibrium constant for the simplest amino acid molecule, glycine, is 2 X 10^{-40} in which J. H. John Peet calculates from this constant that only one glycine molecule would be found in 10,000 liters. By the way, the basic chemical reaction for making glycine is

$$2CH_4 + NH_3 + 2H_2O \leftrightarrow H_2NCH_2COOH + 5H_2$$

As we can see in this chemical equation, the number of molecules to make one glycine are two methane molecules (CH_4), one ammonia (NH_3), and two water (H_2O), yielding one glycine and five hydrogens (H_2). The equilibrium constant is basically the number of product molecules divided by the number of reactant molecules. With the equilibrium constant being ridiculously small (0. with 40 zeros after the decimal point before the 2), the probability of all the gaseous molecules of methane, ammonia, and water colliding together with the right positions and the required energy simultaneously at the right time with intermediates formed to make glycine is extremely low.

To get a grasp on the extremely small number of glycine molecules formed, let us use an ordinary marble, around a half inch in diameter, to represent each of the molecules involved in this reaction. Taking a marble to represent each of the starting molecules (methane, ammonia, and water), how many marbles are needed to yield a single glycine marble and the five hydrogen marbles? I started with a bathtub for holding the marbles and considered how big the bathtub would be to contain all the reactant marbles that would be needed to produce one glycine marble. I quickly realized the bathtub was too small to even come close to containing the marbles representing the reactant marbles needed.

I finally ended up with a bathtub the size of Texas. And even with the bathtub the size of Texas, the height of the tub would reach all the way to the moon! In other words, if you had a bathtub whose size is the area of Texas with a height to the moon, filled it with white marbles that represented all the reactant molecules, and then put one red marble into the tub to represent the glycine molecule, you have a good sense of how low the yield would be.

With this low of a yield, how in the world could glycine have been detected? The answer lies in how the experiment was designed and the related apparatus constructed. An important and necessary part of the apparatus in order to possibly see any successfully synthesized glycine is the cold trap. The trap removes water and water-soluble products from the various products produced, including only 16 of the 20 amino acids necessary for life. It is obvious that intelligence was required to set up the Miller experimental apparatus with the necessary trap to obtain the end result they sought. Traps do not appear out of thin air.

Another major flaw in the attempt to simulate the early earth conditions in the designed glass apparatus was the selection of the starting gases to inject into the closed glass system. Our current atmosphere contains around 78% nitrogen and 21% oxygen. You will note neither of these gases were selected as gases to put into the apparatus. The reason they did not use nitrogen and in particular oxygen is clear. The presence of oxygen is detrimental to the formation of amino acids. Since oxygen is very reactive, the amino acids molecules and even some of the other initial reactant molecules would have been oxidized and changed to an oxide that would be useless for the making of amino acids.

This is analogous to an iron nail being left out and exposed to oxygen so it will rust or iron oxide formed, a much more stable and thermodynamically favored compound when compared to the elemental iron and atmospheric oxygen. Another case in point would be if a glycine molecule was placed in the presence of oxygen, the following decay-type reaction would occur:

$$H_2NCH_2COOH + 3/2 O_2 \leftrightarrow 2CO_2 + H_2O + NH_3$$

	Glycine	Oxygen	Carbon Dioxide	Water	Ammonia
Heat of Formation (KJ/mole)	-528	0	-393.5	-285	-45.9

Carbon dioxide, water, and ammonia are typical products of decay. To determine which is energetically more favorable, we should take the sum of the products' heat of formations, subtract the sum of the reactants heats of formation, multiply each heat of formation with the corresponding coefficient in the chemical equation resulting in a net change of this decay reaction of -589.9 KJ. This would indicate the products of decay are far more stable than the reactant glycine when in an atmosphere containing oxygen.

The arguments presented to counter the destructive presence of oxygen is that the early earth atmosphere was somehow void of oxygen, with the current level somehow showing up much later through vast amounts of oxygen injection created by volcanic gases and photosynthesis of water. The formation of free oxygen from hot volcanic gas is not reasonable since all the free oxygen would readily react with other elements present. And the photolysis of just water from sunlight (another explanation presented of how oxygen came into being in the atmosphere) does not happen to any great extent. The irony is that our source of the replenishing of oxygen gas in our atmosphere is from photosynthesis through a complex process that takes place in living plants composed of various components, cells, and vital plant parts.

The formation of the amino acids for life is the first step in the subsequent millions of steps needed to obtain a living being that possesses the vast numbers of crucial components such

$$HOOC\text{-}CHR^1\text{-}NH_2 + HOOC\text{-}CHR^2\text{-}NH_2$$

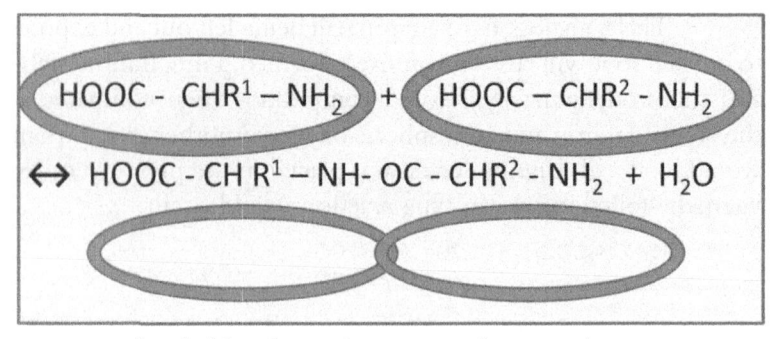

$$\leftrightarrow HOOC\text{-}CH\,R^1\text{-}NH\text{-}OC\text{-}CHR^2\text{-}NH_2 + H_2O$$

as organs, glands, blood vessels, sensory elements (eyes, ears, nose, nerves for sense of feel), nervous system, and skeletal structure, just to name a few. We are only looking at the first few steps and the huge hurdles that exist to determine if these steps can even successfully and actually happen. With the incalculable number of steps required to obtain life as a backdrop to our thinking, what would be the second step needed to take place if we agreed we could create the essential 20 different amino acids? That second step would be the creation or formation of proteins, the building blocks for building a living cell. So how is a protein created from amino acids? To get an understanding of the mechanism of protein formation in this theorized process, we need to look at the central chemical reaction depicted by the formula above.

Two amino acid molecules can join together as depicted above. To accomplish the joining of these two molecules, the H and O from the acid part of one amino acid molecule combines with the H from the amino part of the other amino acid molecule to bond together as one new larger molecule and water (H_2O). To better visualize this linking, I have drawn loops or links around the starting molecules and then show the two links hooked together as the beginning of the formation of an amino acid-based chain. Thus, you can visualize a chain with each link representing a specific amino acid molecule.

The next question to ask is how many of these chain links need to be hooked together to form a protein molecule? The shortest protein contains around 50 and the longest has 2,000 of these links. The links in this chain can only be taken from the group of 20 essential protein building amino acid molecules. The problem is that many other amino acids outside of this group of 20 have been shown to be made in these prebiotic simulation experiments. Thaxton (p. 53) lists 26 non-protein building amino acid molecules that can also react and take the place of an essential protein building amino acid molecule, ruining the successful formation of an allowed protein chain. The odds of this happening is extremely high since the 26 non-protein building amino acid molecules outnumber the protein building amino acid molecules.

Another complication to the formation of a protein is that a protein does not contain just one type of amino acid. Rather, they are composed of a variation of some 20 amino acids, all needing to be linked together in a specific order. The specific sequence of the 20 different amino acid monomer links used and the length of the chain determine the use of the protein in the construction of specific cells and tissue. The human body has at least 20,000 different types of proteins (Omenn, G. S. et al., 2016). To assume that somehow 20,000 different proteins with amino acid chain link lengths of 50 to 2,000 that had the amino acids monomers (out of the 20 possible amino acid molecules) placed in the right order and were somehow randomly formed is utterly beyond comprehension (Wilson, 2002, p. 341).

Some may argue that the process of building proteins through random processes took place over millions of years, one step at a time. The problem is the overwhelming uphill battle in terms of the decrease of entropy that occurs during the process. As stated earlier, entropy is a state function where the final entropy change that occurs from a beginning to an ending state does not depend on the path taken. In this case the loss of entropy, or the tremendous gain in system order that occurs, would be astronomical—requiring tremendous amounts of ordered energy to offset the decrease in entropy.

This issue of getting the 20 possible amino acids to link up in the proper order was actually highlighted in a textbook (Wilson, 2002, p. 344). The authors calculated the total number of combinations using the 20 different amino acid links for an average protein containing 300 amino acid links. The total number of combinations is 10^{390}, which is one followed by 390 zeros.

The odds of getting the one of the 20,000 different acceptable protein chains with amino acids in correct order would be 20,000 to 10^{390} or one chance in 10^{385}. What is truly even more amazing is that in the next paragraph, after highlighting these insurmountable odds, the authors marvel at the natural selection process that had to occur over the long trial-and-error process needed to eliminate the non-usable proteins, or 10^{385} (i.e., one followed by 385 zeros) non-usable proteins.

This tells me a few things. First, they are so far removed from the basic fundamental Second Law of Thermodynamics and its understanding that they cannot see how contrary their thesis is regarding this fundamental entropy law. They have so bought into the evolutionary theory that they cannot step away from its premises and actually see how untenable and unrealistic the idea is of the proteins simply coming into existence on their own. It is a scientific impossibility for this to occur by pure chance.

Another huge hurdle to the formation of any of these protein chains from a proposed prebiotic synthesis as described above is in regard to the observation that all living things have proteins composed of amino acids that are all left-handed molecules. Nineteen of the geometric structures of the essential amino acids are asymmetric, leaving the simplest amino acid, glycine, as the only symmetrically structured molecule. These 19 molecules actually have two structures or formations even though they are identical in the number and types of atoms that make up the molecule. The energy content, properties, and reactivity of these two very slightly different molecular structures are virtually identical. In essence, their molecular structure and shape are the mirror image of the other.

Another way of looking at this is to compare your right with your left hand. They are identical in every way with the same number and type of fingers, nails, joints, etc. At the same time, they are not identical since you cannot overlap one on top of the other, taking both hands palms down and superimposing them. Instead, the left hand is a mirror image of the right hand. You can bring your hands together, palm to palm and fingerprint to fingerprint which shows the mirror image. It is this idea of visualizing these two different molecule structures in terms of human hands where the term right- and left-handed molecules label is derived. It is curious that all the amino acids (except glycine which is symmetrical and therefore does not have a mirror imaged structure) used to construct the different proteins are left-handed.

That brings us to the next problem in the synthesis of amino acids from prebiotic elements (i.e., ammonia (NH_3), water vapor (H_2O), methane (CH_4), carbon monoxide (CO), and hydrogen(H_2). The end products from these synthesis reactions, as was also seen in the Miller-type experiments, resulted in 50% left-handed molecules and 50% right-handed molecules. This greatly reduces the odds of the random assembling of protein chains containing all left-handed molecules when the source of amino acids to build these chains from 50 to 2,000 links long consists of equal amounts of right- and left-handed molecules.

It needs to be restated that there is virtually no difference in the energy content and properties of right- and left-handed molecules so there is no advantage in selecting one over the other when building an amino acid chain. It would be like flipping a coin 50 or 2,000 times to replicate the chance of selecting the next right- or left-hand molecule to add the next link in the protein chain. In other words, one would have to flip the coin and get heads 50 or 2,000 times in a row—a truly impossible feat.

We have only looked at a singular type of protein made. As stated above, there are 20,000 different types of proteins needed to construct the many different biological components of a cell and other tissue types. It requires different proteins to make the cells in the eye, skin, muscle, blood vessel, blood cells, various membranes, nose, bone marrow, and nerves. A good question to ask next would be how many proteins are found in a typical cell in the body. It would be quite a feat to number and catalogue the total number of the 20,000 different protein chains needed to construct a single cell. With the joint effort of many researchers, that number was recently reported in *Science Daily* (2018). The total number of proteins is around 42 million. Keep in mind that is just for one cell.

Chapter Nine

A Tapestry

When one tends to see such huge numbers as we discussed in the previous chapter, the mind tends to go numb and cannot really grasp the size of this number. To better grasp the size of 42 million proteins per human cell, I will use a model to represent proteins that make up or compose a cell. The model selected, strangely enough, is a tapestry weaving. For those unfamiliar with tapestries, they are amazing works of art made on huge weaving looms. The construction of a tapestry begins with hundreds of non-colored threads that are stretched parallel in a vertical position, like strings on a harp that are pushed closely to each other. The artistic exercise in a tapestry creation is the careful weaving of just the right color threads horizontally, in and out, and between the non-colored vertical threads.

One of the oldest and most established factories for tapestry art in the world is in France and is named Gobellins.[1] Located in the suburbs of Paris, it was established in the 15th century and was the official state-run tapestry manufacturer, making tapestries commissioned by French royalty. The precision and care needed to lay out and setup the materials and then perform the actual hand weaving that will portray a picture, scene, or story are absolutely amazing.

A YouTube video[2] shows a weaver having a sketch of an intended scene with markings drawn on the vertical non-color threads to help ascertain reference points, and a large mirror

---
[1] see https://nazmiyalantiquerugs.com/blog/gobelin-tapestry/
[2] See https://youtu.be/jIbu-dJuEh0

positioned on the front side facing the tapestry being made. This mirror setup is so the weavers working behind the tapestry can see what they are weaving and compare it to the sketch of the scene. The number of different colored threads is generally less than 10. I hope from this brief description you come away with a deeper appreciation of the craft of tapestry.

To use a tapestry as a model for the 42 million proteins that make up a cell, I am going to use the colored horizontal threads to represent the protein molecule chains which would be 42 million protein chains, the number of protein molecules that have been estimated to make up a cell. I will not get into where the threads of different colors came from. One could look at the intricacies of thread using cotton and wool harvested from a farm, or synthetic threads from a chemical processing plant and the dyeing of the thread to achieve just the right colors. The different colors will represent the different types of proteins in a cell which as you will recall from above is around 20,000.

So right away it can be seen this tapestry model is pushed beyond its limits where the different number of colors are typically 10 or less for a tapestry, far less than the 20,000 of proteins found in a cell. From a supplier of tapestry wool on the internet site www.dmc.com, there are 390 colors of wool available. In this model, the number of colors needed to accurately depict the different proteins in a cell would be 2,000 times more colors than the 10 colors that are typically used in a tapestry. The hues and colors required would rival even the best flat-screen colored television.

As a comparison to the model of a tapestry representing the number of total proteins in a typical human cell, I will refer to the work entitled "Adoration of the Magi" which was created by the famous tapestry factory Morris & Company[3] founded by William Morris in London. When Morris started making tapestries by experimenting with a loom he had in his bedroom, he worked on it for countless hours after work. Then he went to France and learned firsthand the techniques for dyeing the

---

[3] https://www.exeter.ox.ac.uk/morris-and-burne-jones-tapestry-to-be-housed-in-museum-quality-display-conditions/

materials used in the weaving. In 1861, he and his partners started a business that would eventually become Morris & Company. All he learned he passed on to his employees.

The Adoration of a Magi was commissioned by Exeter College, Oxford in 1886.[4] The artist who painted the watercolor used as a template to make the actual tapestry was Edward Burne-Jones. It took three weavers two years to complete. The tapestry, shown in Figure 5, depicts the Nativity scene of the visit of the three kings after the birth of Jesus. From the photo of the tapestry, you can see the tremendous amount of detail and the intricacies of the weave. Since this book is published in black and white, to gain a deeper appreciation of the beauty and complexity of this tapestry, I encourage you to find a colored depiction of "The Adoration of a Magi" on the Internet. It is amazing.

Figure 5. The Adoration of the Magi, copy woven in 1894 by the Morris's Merton Abbey Mills

The dimensions of this tapestry are 101 inches by 151 inches or 258 cm by 384 cm. Now that we have an idea of the amount of organization, planning, set up time and work, and the vast amount of ordered energy expended to make a detailed

---

[4] https://www.exeter.ox.ac.uk/morris-and-burne-jones-tapestry-to-be-housed-in-museum-quality-display-conditions/

tapestry such as the "Adoration of the Magi," I will now proceed to lay out the tapestry model for all the proteins that exist in a single living cell, using the conservative assumption that the length of one amino acid link in a protein is comparable to its width.

The Adoration tapestry used 12 woven threads per vertical inch: 12 rows of thread woven perpendicular to and between the parallel vertical non-colored threads for every vertical inch. The 12 threads per inch translate to about 2 mm width or diameter of colored thread or yarn. To make the calculations easier, I will suggest thread or yarn that is one mm wide or one mm in diameter. This translates to 24 threads per vertical inch, doubling the resolution, which will mean a sharper image woven into the tapestry, while at the same time doubling the time required to make the tapestry.

If we take as an average a protein being 100 amino acid links long, the diameter of the thread set at 1 mm would yield the length of thread to represent this 100-amino-acid-linked chain would be 100 mm or 0.1 meters. This would keep the same scale for both the vertical and horizontal dimensions, similar to a road map with a mileage scale that can not only be applied to the east/west roads but also to the north/south roads. As stated above, there are 42 million proteins in a human cell which could be translated into a total length of different colored threads.

Using the above selected scale of dimensions, if we lined up all the threads, which represent the cumulative length of all proteins in a cell, in one continuous row, it would stretch out to a total length of 4,200 kilometers or 2,600 miles of colored thread. What would the size be of a square tapestry using these defined scaling factors? It would be an unbelievable 25-mile by 25-mile square or 625 square miles. The *Adoration of the Magi* occupies 10.5 square feet. So how much bigger is the tapestry modeling the number of proteins in a cell? It would be about one billion times larger, meaning it would take three billion weavers two years to make the proteins into a cell tapestry model. That is an astronomical amount of ordered energy needed to complete this tapestry, not to mention the use of 20,000 colors that would represent the 20,000 different proteins that make up a cell.

Let me add one short clarification concerning these proteins. In real life, they do not naturally stretch out in a straight line, but rather fold together into a tight three-dimensional configuration. That being said, the tapestry analogy still holds since it is a comparison of how the 42 million total proteins consisting of 20,000 different types of proteins are fitted together to yield a living cell. This is similar to how a tapestry composed of thousands upon thousands of different colored threads are fit together to yield a beautifully created scene or picture.

LIVING CELLS

In actuality, the living cell is much more complex than the tapestry due to all the different active processes taking place for the cell to be living, such as respiration, nutrient absorption, expelling of wastes, cell division, and cell protection. I once heard the complexity of processes and actions that regularly occur in a living cell rival all the actions and events that go on with all the human activity in New York City. Given that 42 million protein molecules are needed to construct a cell, with each of the 20,000 different proteins having its own function and comparing this to the population of New York City of around 8.5 million people, the comparison of a cell's complex activities to that of all that goes on in the city is a reasonable one. Multiply that one cell by the 37 trillion different cells that compose a human body, and the level of complexity explodes beyond belief with no easy comparison to aid in comprehension.

This tapestry analogy should help illustrate the insurmountable thermodynamic mountain that would need to be overcome. The mountain in this case being the immense amount of ordered energy required to overcome the Second Law of Thermodynamics' natural tendency of disorder. A tremendous amount of the increase in energy content and decrease in entropy occur from turning simple molecules like water, ammonia, methane, hydrogen, and carbon dioxide into a living cell occurs. Currently, the only way this is accomplished in day-to-day life and living is through already existing complex living systems. I repeat that once a living being dies, all the processes that yielded the complex and ordered cells and systems with inherently higher energy contents abruptly stop. The existing complex biomolecules within this once living system then begin to break down to their more thermodynamically favorable states and smaller molecules. Without a living system, these complex biomolecules cannot be made and sustained. The complex order and operation of a cell are phenomenal.

DNA

The largest biomolecule that exists in every living cell, both in animals and plants, is the most amazing biomolecule studied of all. It is ironic that the proposed structure and identification of this remarkable biomolecule occurred in 1953, the same year Miller presented data on the prebiotic simulation and experimentation of obtaining amino acids from select gases and an electric spark within a closed system. Over the years, Miller's experiment, though talked about in classrooms with much hype, has not really yielded any great breakthroughs. The biomolecule I will be discussing is quite the opposite. Literally thousands of institutions have studied and catalogued its structure and the various modifications observed between different species and the variants discovered within a species. The variants found and studied have provided significant advancements and understanding in the field of medicine.

If you have not already guessed what this gigantic biomolecule is, it is deoxyribonucleic acid, better known as DNA. Even though DNA was first isolated by Friedrich Miescher in 1869, its specific structure was not identified until 1953 by Frances Crick and James Watson at the University of Cambridge. By using many x-ray diffraction measurements, Crick and Watson deciphered these complex measurements and built a model of DNA which is commonly referred to as a double helix structure. In plain language, it looks like a twisted rope ladder.

What is truly amazing about the DNA biomolecule is that it holds all the codes and information necessary for replicating proteins used to construct materials within a cell. What is astonishing is that not only does this information in DNA determine what species it will be (a human being, a horse, a dog, a snake, a fish, a frog, an oak tree, a dandelion, a honeybee), but it also determines the different characteristics the specific living being has within its species, such as eye and hair color, physique, plant-fruit-seed size, breed type, etc.). Another piece of information obtained from the DNA code is actually what a cell's function will be within the living organism. Will the cell be used to be part of the eye, or be a skin cell, a blood cell, a liver cell? When pondering all this information encoded in DNA, one has to admit this is truly quite a remarkable biomolecule.

The molecular structure of a very small portion of the DNA molecule is shown in Figure 6. The atoms used to build this complex macro-biomolecule are hydrogen, oxygen, nitrogen, carbon, and phosphorus. The figure shows the location of these atoms and how they are bonded or linked to each other. I will simplify the explanation of this DNA structure and its function. An overall observation of the DNA molecular model reveals that it is basically a double helix structure or the twisted rope ladder. There are two basic parts to this ladder—the side rails where the phosphorus atoms are located and the ladder rungs. The rungs are what contain all the coding or information. The side rails simply provide the structure to hold the rungs. The rungs are made up of only two types.

Each rung has two halves that are weakly bonded at the middle of the rung by what is called hydrogen bonding, a bond that can easily be broken and yet put back together. That bond is like a magnet and a piece of iron, which will stick to each other but can be pulled apart with little effort. These easily broken bonds at the middle of each rung are essential to a DNA's mission of getting information out and reproducing itself. There are only two pairs of molecules used to make up the rungs, which are thymine (T) that is weakly bonded through two hydrogen bonds to adenine (A) to make up one rung and cytosine (C) weakly bonded through three hydrogen bonds to guanine (G) to make the other rung. The sequence of how these two different rungs is lined up along the ladder is how the information is stored.

There are actually four ways to orient these two differently paired rungs or base pairs as they are called. Using the letter assigned to each molecule portion, the four-rung orientations are:

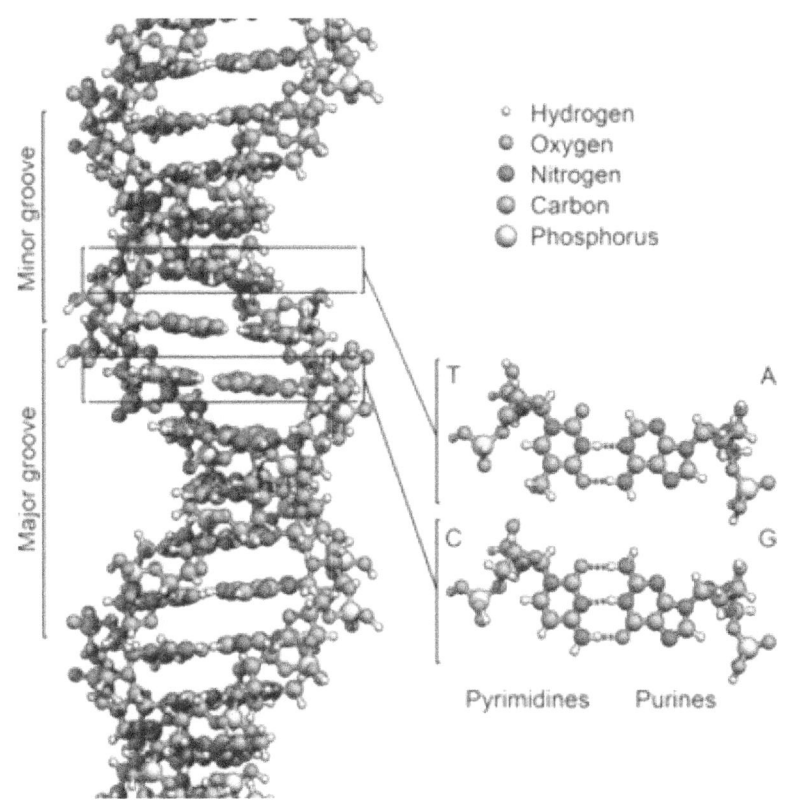

Figure 6. A drawing of a small section of the DNA molecule

A-T, G-C, T-A, and C-G. Not only do the two actual molecule pairs contain the information but by flipping each pair 180 degrees, the possible information to code the DNA rung sequence doubles. In short, there are only four different combinations of ladder rungs (base pairs) that can be placed in a sequence along the DNA ladder which is what provides the actual information encoded into the DNA supermolecule.

So where is this DNA located? It is located in the nucleus of a cell. It is sort of the brains of the cell, containing all the necessary information that determines the cell's makeup and purpose. For humans there are 23 pairs of DNA macromolecules, which make up the chromosomes of a cell. A gene is a term widely used when talking about traits of living beings. Genetics is the study of these genes and how these genes dictate what a living being is. A gene is actually determined by the sequence of based pairs or ladder rungs within a DNA strand or ladder, and usually a few thousand base pairs or rungs compose a single gene.

So how does a DNA macromolecule function to express the information it has contained within? There are two main actions of a DNA molecule, both of which entail the splitting the ladder rungs down the middle of sections of the DNA molecule along the hydrogen bonding sites and exposing them. It is the sequence of these bondable hydrogen sites along the DNA chain that acts like a template for the eventual synthesis of other macromolecules by individual monomer molecules that move to these open bond sites and are weakly bonded to the exposed DNA hydrogen bonding sites.

Once these monomer molecules are lined up along the exposed DNA hydrogen bonding sites, they then bond to each other forming a custom-designed macromolecule matching the template of the DNA molecule. This action, when accomplished along the whole length of the DNA molecule, is the replication of an identical DNA molecule that will be used to create another cell. This cell replication is necessary for cell reproduction as part of the growth of the living being or the replacement of older cells that will die and need to be replaced. The other action that uses the same replication process with only small sections of the DNA molecule is the production of a multitude of different proteins for use in the construction of cells and biomaterials necessary for the living being.

As you will recall from above, for human beings there are 20,000 different proteins needed within a cell for a total of 42 million proteins required to build a single cell. To accomplish this, something called messenger RNA molecules are formed from individual DNA hydrogen bonding sites, thus taking this site's info for use in a process called information transcription. This small messenger RNA molecule containing the DNA information is free to move to outside the nucleus into the cytoplasm region (cytoplasm is the liquid contained within the cell walls with the nucleus containing the DNA at the center of the cell surrounded by this liquid cytoplasm). The liquid cytoplasm allows for easy movement of RNA and other pre-protein molecules. All the messenger RNAs then act as a template for the building of longer chains of specific proteins. This part of the process is called translation of the DNA information. Figure 7 provides a simple schematic of this transcription and translation processes.

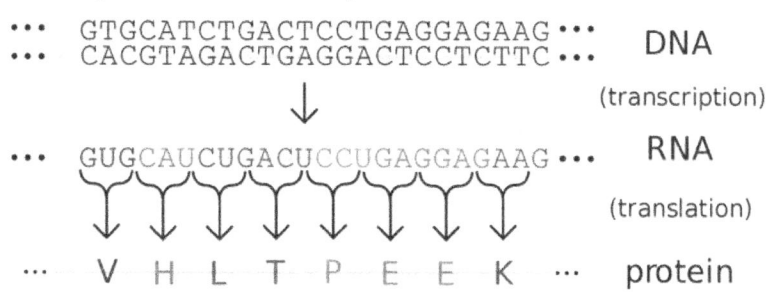

```
···  GTGCATCTGACTCCTGAGGAGAAG ···    DNA
···  CACGTAGACTGAGGACTCCTCTTC ···
                    ↓                (transcription)
···  GUGCAUCUGACUCCUGAGGAGAAG ···    RNA
      ⌣  ⌣  ⌣  ⌣  ⌣  ⌣  ⌣  ⌣         (translation)
      ↓  ↓  ↓  ↓  ↓  ↓  ↓  ↓
···   V  H  L  T  P  E  E  K   ···   protein
```

Figure 7. A simple schematic of the transcription and translation process

What has been provided above is an oversimplification of what actually happens within a cell in regard to DNA's function and purpose. The necessary folding of proteins with helper molecules and enzymes, how the process knows when to start and end the building of a macromolecule, along with many other specifics reveal more levels of complexity of order being carried out within all the cells of a living being. What I have presented should give you an appreciation of the astronomical extent of order going on continually at a microsecond pace within the billions of cells within your body.

As I have clearly stated, all this can only go on within a living being that takes in energy and then like a machine puts the energy to use in an ordered way to produce the macromolecules possessing tremendous amounts of order and loss of entropy. If that living being dies, the whole ordered process stops and the increase in entropy begins according to the Second Law of Thermodynamics. Even while a being is alive, the fight against entropy is still obvious. The increase in entropy is why aging occurs. Over the course of time, the order that exists with DNA breaks down causing the body not to function as it did when younger.

Also, over the course of human history, outside agents such as light, chemical exposure, and other factors attack and mutate DNA molecules to a point that the defect can be encoded in DNA and therefore replicates the defects in future generations. This is the source of heredity-linked illnesses and defects. We all have these defective genes within our DNA codes. This therefore is the reason for restrictions in marriages between a man and woman who are closely blood related. This is also the same reasoning for avoiding the breeding of animals that are closely linked by bloodline. For the vast majority of these defects coded into DNA, they have a recessive trait and will only be realized when both the mother and father have the same recessive defect in their genes.

When thinking about the overall human history in regard to DNA makeup and the changes in DNA over time with respect to the Second Law of Thermodynamics, it makes more sense to consider that at one time the whole DNA structure existed in a pure form without any defects and mutations within this macromolecule. This would fit with the idea of there being an Adam and Eve as the precursors to the human race, both possessing DNA that is perfect without any defects. Then over time, this pure form of DNA changes, becoming slightly defective, resulting in heredity linked problems and diseases or hidden defective traits that are realized in future generations.

The DNA macromolecule exists within the nucleus, folded and twisted into a small space. I remember years ago reading about DNA to better understand this amazing molecule and discovering if one took the DNA strands within the nucleus of a human cell and stretched them into a straight line, it would stretch to six feet. Being familiar with atomic distances, I was truly amazed at the length of DNA. To gain a better appreciation of this length, I decided I would make a mental model of this by using a rope ladder with rungs twelve inches apart. The distance between consecutive base pairs in DNA is stated to be 3.4 angstroms. Using this information one can calculate the number of rungs needed in the rope ladder model and then calculate the model's length. When this calculation is done, the length of a rope ladder with one-foot distance between rungs would need to be one million miles long, which would stretch 40 times around the earth's equator.

When considering that the sequential order of base pairs along the ladder needs to be in a defined and specific order, far removed from any random selected order, a great deal of intelligence is required to get this delicate balance of sequential order right. If the order of the four possible base pair sequence is off by even a few rungs, this could lead to disaster for the living being and its non-existence. In other words, one cannot just randomly place the order of the rungs on the DNA ladder with the four different possible base pair arrangement (like A-T, G-C, T-A, and C-G). What greatly magnifies this issue of intelligence rather than randomness required to build the DNA chain is when one considers that each species has its own set of DNA sequences. Of course, there is a variation within a species' DNA make-up to account for the various different physical properties, but those numbers of DNA-strand sequences to account for these variations is miniscule when compared to the total number of overall possibilities of base pair sequences on this extremely long ladder.

What ultimately defines a species is its DNA make-up as DNA macromolecules group together in pairs to create chromosomes of which each species has its own unique number, sequence of the rungs order, and length. A human has 23 pairs of chromosomes, a fruit fly has 4 pairs of chromosomes, a rice plant has 12 pairs, and dogs have 39 pairs. It is often stated that there is only a 2% difference in the chromosomes between humans and chimpanzees. In reality, the 2% difference in terms of the difference in the base pair sequence is quite significant. Using the same rope ladder model outlined above that would stretch 40 times around the earth to model a human's total DNA length, the 2% difference in the model would be a section of DNA that would circle the globe one time. Keep in mind that the rope model was calculated with the basis of one foot between rungs, with each rung representing a base pair.

Humans and chimpanzees are two different species defined by two different DNA compositions. I listed three other diverse species: fruit fly, rice, and dog. This brings up the question of exactly how many species exist on earth. Liz Osborn compiled a list of species from many sources and listed them in Currentresults.com.[5]

Table 1 Total Number of Species Within Classes	
Vertebrates	Number of Species Within Class
Mammals	5,500
Birds	10,00
Reptiles	10,000
Amphibians	15,000
Fishes	40,000
Total	80,500
Invertebrates	
Insects	5,000,000
Arachnids	600,000
Mollusks	200,000
Crustaceans	14,000
Echinoderms	14,000
Others	791,830
Total	**6,619,830**
Plants	
Flowering Plants	352,000
Conifers	1,050
Ferns & Horsetails	15,000
Mosses	22,750
Total	**390,800**
Total of All Species	5,500

[5] Please see the disclaimer at www.currentresults.com/About/termsuse.php.

This list of the number of the currently known species is mind boggling, resulting in a total tabulated number of species of over seven million. The effort to identify, catalogue, and compile this table was a herculean effort with contributions from thousands of people. What is even more amazing is when one considers the DNA of each of these species, which is actually the underlying factor in defining the species. Now that you have a grasp of the extreme complex order of the DNA for a person as demonstrated as a rope ladder wrapping around the earth 40 times using my one-foot-apart ladder rung example, you should appreciate the astronomical magnitude of the overall order required in the above table of total species. It is beyond comprehension. There is no way the above combined order of all the species listed in the table above could have come about through a long series of random acts. The loss of entropy from the creation of the species represented in the above table alone is incalculable. It is obvious that some sort of intelligence far beyond our comprehension had to bring about the order of all the DNA molecules of all the seven-million-plus species that exist on the earth.

Chapter Ten

The Human Body

One last topic we will explore also represents the highest level of complexity in a human being: the physiological systems of a person. The table below lists these different physiological systems.

TABLE 2

List of the Different Systems in the body:

Skeletal System	Muscular System
Respiratory System	Urinary System
Nervous System	Endocrine System
Circulatory System	Digestive System
Reproductive System	Lymphatic System
Integumentary System	

As stated above, one part of the information coded within the DNA molecule defines the purpose and function a cell has. That function fits into one of these eleven physiological systems. We will briefly look at a few of these systems in order to appreciate the order and complexity that exists even within these systems themselves. To enhance your admiration for the beauty of the design of the systems and the degree of system engineering represented in each, I will compare them with some of the features of those that engineers build within man-made machines.

Skeletal System

Let's first look at the skeletal system that provides the

framework and structural support for the entire body. Newborns have 270 bones with some bones eventually fusing together during the baby's growth to yield 206 bones found in an adult. The brilliant reason for this fusing of the bones after birth is that it allows the baby's body to be more pliable to allow easier transport down the birth canal. After birth, the more flexible multi-piece bones and cartilage fuse together forming the necessary rigid structure needed for a more mobile, structurally-protected body able to walk and be less vulnerable in its environment.

The amazingly designed ball-in-socket structures provide fantastic range and degrees of motion found in fingers, hands, wrists, forearm, elbow, shoulder, and vertebrae. All these joints work together seamlessly to carry out a desired activity. Think about the collective motion of all the joints in the arm needed to swing a hammer, pound a nail, or throw a baseball. The hand itself needs to have the flexibility to conform to a shape that firmly holds on to the hammer but yet is able to have a tight grip on a round baseball with the motion of joints from the wrist, elbow, and shoulders, all working together to hammer a nail or throw a baseball. This represents an amazing piece of engineering.

This type of functionality for accomplishing specific and variant mechanical tasks requiring a full range of motion to any degree is nearly impossible to mimic in robots designed by whole teams of engineers. The last detail and one of the most important aspects of skeletal bones is their bone marrow for it is there where vitally needed blood cells are manufactured to continually supply the blood cells needed for the circulatory system. The marrow is perfectly protected within the bone so this can take place.

Muscles

Let's look next at the muscular system which is nothing but sets of complex but efficient actuator systems. Each muscle group is composed of a pair of bilateral muscles, one providing motion in one direction while the other counters the first by pulling the skeletal component in the opposite direction. This ingenious pairing system in conjunction with the bone's ball-in-joint hinging mechanism allows for precise control and positioning in virtually all degrees of motion.

The activity of these muscle pairings can easily be experienced using your arm muscles, namely your biceps and triceps. These are the pair of muscles in your upper arm that enable you to move and exactly position your arm up and down. The biceps is the arm muscle people use to show off how muscular they are. When you take your arm while sitting at a table and place it with your palm facing up and touching under the tabletop and then pulling your palm against the bottom of the table and lifting it up, you can feel with your other hand your bicep contracting and hardening. This bicep muscle is contracting and trying to shorten its length to help the arm pull up while pivoting at the elbow.

The triceps, the muscle that opposes the bicep, can be seen in action by using the same scenario. This time, place your hand on top of the table with the back of your hand resting on the table. Push down on the table with the back of your hand. This is the action opposite to the action done under the table to illustrate a bicep. With your other hand you can feel and see that the bicep is softer and flabby, while the opposing triceps on the back of the arm is stiff.

In order for the body to accomplish all its movements of its skeletal components, 320 pairs of muscles are required. All these muscle pairs need to coordinate with each other to carry out the tasks we take for granted such as walking, running, swimming, writing, painting, talking, chewing, blinking, moving our eyes, shaking our head yes or no, wiggling our toes, clapping our hands, and swinging a golf club or tennis racket—and those are only to name a few. Any motion of the body involves many of these 320 muscle pairs.

NERVES

The physiological system to discuss next is the nervous system, the communication link between the brain and the rest of the body. The nerves have two main functions. The first is to communicate with the muscles with one of two purposes. The first purpose is to carry out the muscle activity performed automatically and without focused or conscious attention. The second purpose is to control voluntary muscle activity of an individual. The desired actions can be started and stopped voluntarily, activities such as walking and writing.

The second function of the nerves is to communicate to the brain the sense of touch experienced on the skin, which can take on a number of dimensions. One dimension is the sense of temperature like when we touch a hot surface or a piece of ice to any part of our body. In addition to sensing temperature changes, these nerve endings can also experience pain. Both of these senses, temperature and pain, are how the brain senses danger at different areas of the body. Part of the horror of leprosy is losing a sense of touch because the disease deadens the nerves so the brain is not able to sense if a dangerous situation is occurring somewhere on the body.

The nerves through the sense of touch also provide feedback to aid in the precise movement of body parts by the bilateral muscles. One simple example of this is the picking up of an egg without squeezing and breaking the egg. To accomplish all the above stated functions, 43 pairs of nerves are needed which have a total length of 45 miles. This is a fantastic system that robotic engineers cannot easily duplicate in any of their creations.

Circulation

Let's look next at the circulatory system, vital to any living system's existence. The human body is estimated to contain around 37 trillion cells. As stated above, these cells have a wide variety of functions but need to carry out their functions in coordination with all the other cells. To carry out these functions, the cells expend energy and need fuel or nutrients to produce the energy or movement needed. One of the circulatory systems main functions is to deliver nutrients and oxygen to each cell. The cell obtains their energy stored in the blood containing nutrients through an oxidation process.

This process creates the lesser energy containing molecules of water, carbon dioxide, and other waste products. The red blood cell contains hemoglobin that loosely complexes with and grabs onto oxygen molecules obtained from the lungs. The red blood cells travel through the circulatory system reaching all the cells to deliver the oxygen and then picks up the carbon dioxide waste to deliver back to the lungs to expel from the body into the air. Besides supplying nutrients to and taking away waste from the cells, the circulatory system is the main defense for cuts or other injuries that occur on the body. It provides platelet cells and proteins called fibrin to wounds to promote clotting and stop bleeding, also supplying white blood cells to fight infection and delivering an oxygen-rich environment to facilitate new cell growth at the wound. The heart pumps an astounding 2,000 gallons of blood per day, which is more than thirty-six 55-gallon drums! What is even more astonishing is the length of all the blood vessels and capillaries to deliver the blood to all the cells is 60,000 miles, which is more than twice the distance around the earth at the equator!

Digestion

The last physiological system to review is the digestive system. This system acts like a power plant or factory for taking in food and breaking it down into a suitable form to obtain the food's stored energy as well as its vitamins and minerals for the cells to use. In order to see how amazing the digestive system is, let us follow the path the food takes starting at the dinner plate and ending with the food's entry into the blood.

The ingested food is actually the fuel used by the cells. Fuel is where energy is chemically stored that powers a machine or engine. In this case, the calorie content of the food is what is used to power the body as it supplies energy for all the cells. First, let's compare the fuel used in man-made machines to that of our body or bio-machine. Most automobiles have engines that use gasoline as the fuel. There are three to four grades or octanes that can be used, which is nothing but a slight variation in the liquid hydrocarbon content. A car engine is picky in regard to the food or chemical energy it will use. You would never pour vegetable oil even though it has good energy content into a gas tank. Doing this would result in an expensive car repair bill. Even diesel, a close relative of gasoline, proves to be disastrous in a gasoline vehicle. Our bodies also require a certain kind of fuel—not just any fuel will do.

But there is an unbelievable number of fuels or foods we can use to supply our bodies. We can power our bodies with pizza or steak, broccoli or corn, bread or spaghetti. You get the idea. This fuel variety for our bodies is much more extravagant than for man-made machines like cars, for which only a few options are acceptable. Now let's track the food through the digestive process.

Once the food leaves the plate, it needs to pass through a number of steps before it is transformed into a form compatible for absorption into the cells. The first step involves the mouth where the food is broken down into smaller pieces so it can easily pass through the throat or esophagus to the stomach. In addition to the chewing action, the saliva glands excrete liquid to aid in the pulverization process. The taste buds not only provide pleasure through taste and texture, providing incentives to eat, but they also alert the body to food spoilage or other harmful toxic items.

Once in the stomach, the pulverized food is chemically broken down even further. The esophageal sphincter, a one-way valve, is positioned at the top of the stomach at the esophagus exit to keep the food in the stomach. Hydrochloric (HCl) acid in the stomach breaks down the food particles into chemical compounds that can be absorbed later in the digestive system. The HCl is generated by parietal cells located in the stomach. The cells use a multistep process of making carbonic acid from water and carbon dioxide. An enzyme takes and converts the carbonic portion of the carbonic acid (leaving the hydrogen ion, the key component of acids) to bicarbonate which is exchanged for chloride ions that pair up with the hydrogen ions which yield hydrochloric acid.

The pH of HCl, or the concentration of hydrogen ions in the stomach, can be from 1.5 to 3.5. The pH of highly concentrated HCl, which is 38% HCl and 62% water is 0. This concentration is so strong it can dissolve skin. The concentration found in the stomach, though not as harmful as concentrated HCl, is strong enough to damage and irritate body tissue. Many have experienced and felt this type of damage when they have heartburn. This occurs when the check valve at the top of the stomach does not properly close and routinely allows stomach liquids containing HCl to travel partway up the esophagus, irritating the walls of the throat.

This begs the question. Why is the stomach wall tissue not damaged by the HCl that it contains and produces? The stomach has a few mechanisms to protect itself from being digested by the

HCl. First, cells in the stomach lining create a protective mucous lining. Also, bicarbonate is also excreted from the stomach lining that neutralizes HCl at the lining. When there is a failure of any of these protective measures, stomach ulcers can occur. After a few hours in the stomach and once the food particles have been broken down by the HCl to yield useful proteins and fats, the digested food travels toward the intestines through the ileum where the digested nutrients are absorbed into intestine walls and passed on to the circulatory system.

The large intestines accept what is left and will absorb any remaining water, leaving behind stool to be excreted. Even through this abbreviated and simplified explanation, you can see that the whole process of extracting the energy from the food requires a complex process and system. This system has all the markings of being designed by some great intelligence that exceeds our engineering capabilities, and for sure could not have come about by random processes of chance.

THE FIVE SENSES

The last set of systems I will mention is the sensory systems of sight, hearing, smell, taste, and touch. They all also possess a highly ordered complexity that is seamlessly integrated with the other physiological systems. Let's only examine sight. When we consider the various photo biochemical reactions involved, the electrical impulses generated, the various subcomponents used to make up the whole sensing function, and the perfect incorporation within the physiological systems, our amazement of a living being's design goes to a new higher level.

Search the Internet and you will gain an even greater appreciation for how the eye works. Some websites even discuss the different biochemicals that are transformed by light, resulting in large electrical charges that are sensed and transmitted by the optical nerve to the brain for interpretation. In addition to the nervous system playing a large part in the sense of sight, the muscular system has an important role in activities from focusing of the image, to eyeball movement to where one wants to see, and to opening and closing the eyelid. The skeletal system provides for the support and protection of the eyeball within the socket of the skull. The circulatory system provides the necessary nutrients and other essential ingredients such as vitamin C and E and zinc.

These few sentences on the function of the eye do little to describe the incredible complexity of the eye's function where thousands of chemical reactions are generating a myriad of electrical signals transmitted over the optical nerve to a section of the brain to interpret the signals—all accomplished in microseconds. To vastly compound this complexity, consider the other five senses, for which each has its own sets of biochemical reactions generating their own signals sent through the nerves to the brain for interpretation. Again, the amount of order and complexity defies the assumption that this was generated by a random process. Instead, it speaks to design and the makings of unbelievably superior engineering.

This concludes the science section of this book. My goal was to show you how unlikely any living thing could have come about by chance. No, those living entities were produced by some amazing, beyond our comprehension, high levels of creativity and engineering. During presentations I have given to groups that include much of the above information, many times I quote a scripture from Psalm 139:14, which seems apropos in light of the extreme complex order shown in living beings: "I praise you because I am fearfully and wonderfully made; your works are wonderful, I know that full well." The *you* in this quote is a reference to God. Also, the word *fearfully* does not mean scary or frightened, particularly since it is paired with the word *wonderfully* in this passage. Instead, it means to be in a heightened state of awe and reverence.

When I look at the

- levels of bio-assembly from the starting point of atoms of hydrogen, ammonia, water, and carbon dioxide; starting with the
- creation of the 20 essential amino acids for which 19 have to be all left-handed molecules; for
- assembly of amino acid molecules to make the 42 million protein molecules made up of 20,000 different types of proteins needed for cell fabrication;

- the synthesis of DNA that contains all the information and coding to define the living being and the function of the cells composing the living being;
- the human DNA model depicted as a rope ladder with one-foot spacings between rungs that would stretch more than 40 times around the world;
- eleven extremely complex physiological systems that work together seamlessly to enable a being to live and function;
- truly amazing functionality of the five senses.

I find it impossible that the over seven million species of animals, plants, and insects each species with their own unique DNA set, came about by chance. We were designed and made by an intelligence that is infinitely superior to anything we can imagine. To think otherwise is to fool oneself and defy the sense of logic and reason as put forth in the above scientifically based treatise.

The Second Law of Thermodynamics points toward the tendency of disorder rather than an entity becoming more ordered. If everything was created at the beginning with perfect DNA, then one would expect over time the DNA would trend toward becoming more disordered through mutations—which is what we see. After looking at even a small amount of scientific information provided above in conjunction with the Second Law of Thermodynamics, it takes much more faith to believe in evolution than to believe that an intelligent being like God was the Creator of it all. I will next proceed to put forth evidence that this God is the God portrayed, loved, served, and worshiped through the Christian faith.

SECTION THREE

The Bible and Religion

Chapter Eleven

Science and the Bible

      This chapter's title affirms two things I hold to be true. One is that the Bible defines who God is and provides an account of what He has done, how He has worked in history with and through people, and the kind of relationship He seeks with people—including how He has and is accomplishing this. Second, the Bible is His authentic Word, the means through which He has communicated to us as He inspired many authors who have written it. Taking this statement further, I also believe it is the only authoritative book or piece of literature that can make these claims. You may be wondering about the veracity of such definitive statements.

      At the beginning of Chapter Six, I made a bold statement that God is the Creator of all things and then I put forth a long science-based treatise backing this claim. In this chapter and section, I will also present sound reasoning for my claims that the Bible is the only source of foundational knowledge concerning God. This idea of Jesus being the revelation of God and the Bible as the singular true source of knowledge and truth concerning God is alluded to in the Bible. A few Bible passages, both written by Paul in letters to his dear friend Timothy, highlight this point:

> This is good, and pleases God our Savior, who wants all men to be saved and to come to a knowledge of

truth. For there is one God and one mediator between God and men (people), the man Christ Jesus, who gave himself as a ransom for all men (people) – the testimony given in its proper time. And for this purpose I was appointed a herald and an apostle – I am telling the truth, I am not lying – and a teacher of the true faith to the Gentiles (1 Timothy 2:4-7).

All scripture [the Bible] is God breathed and is useful for teaching, rebuking, correcting and training in righteousness (2 Timothy 3:16).

Jesus also stated emphatically that He had the correct teaching and is the source of truth. John 9:31 in the New Testament states, "If you hold to my teaching, you are really my disciples. Then you will know the truth and the truth will set you free." Later in John 14:6, Jesus said, "I am the way, the truth and the life. No one comes to the Father (God) except through me." It is curious that as one examines history, there are places and times when mankind has exerted great effort and energy to eradicate the idea of God, and in particular the Bible and the God it describes. The more recent societies and governments that come to mind are those based in Communism as espoused by its founders Lenin, Marx and Mao. They saw religion as harmful to people and their cause, calling religion the opiate of the people. Their belief was an atheistic government and society to which people were accountable was preferable to any accountability to God and His standards.

It is interesting that at the start of these Communist revolutions, a main focus was not only to discourage and ultimately outlaw the religious practices and thinking but also to replace them with a whole new philosophy of life. This paradigm shift in religious thought was usually extreme considering the previously held religious beliefs in places like eastern Europe where Catholic and the Orthodox churches had a strong established presence.

In addition to forbidding Bibles and non-state sponsored church services, these regimes also introduced a change in the educational curriculum. Not only did they forbid the practice of religion, but they also removed the idea of God being the Creator. The solution to this was, at the time, a new craze in the science world called Darwinism. Darwin had published his treatise stating that since there are similarities found in living organisms, with some apparently being more simplistic and others more advanced, then it was obvious that an evolution of species occurred through random chance and natural selection. This provided an alternative theory as to how living things and more specifically people came into existence, thus removing a key reason or proof for the existence of God. The teaching of evolution as a proven science within their state-run schools was the linchpin of their atheistic system.

It is sad to see that there are currently many schools and universities within the Western world that also promote the same ideas and philosophies. This is ironic since almost all of the first universities founded in America, places like Harvard, Princeton, Yale, Columbia, Dartmouth, and William and Mary, affirmed in their founding charters the importance of the Bible as the basis of teaching and training carried out at that institution of learning. One of their main original objectives was also to train future ministers and missionaries.

The first reading textbook in the American colonies, called the *New England Primer*, was first published in 1688 and used for 150 years to teach the alphabet to elementary students. It associated letters of the alphabet with concepts from the Bible and/or Christian faith. As a matter of fact, one of the main reasons for establishing an educational system in the United States was to enable people to read so they could read and learn from the Bible.

So there has and is an ongoing resistance in the world to the truth of who God is and His desired relationship with mankind. A fundamental part of this resistance is the attempt to discredit the Bible with arguments that the Bible is archaic literature with no relevance to today's world because it is filled with inaccuracies and inconsistencies. There have also been attempts to water down and modify the core Christian beliefs to be more inclusive of other religions and philosophies. This idea is also antithetical to the Christian faith as espoused in the Bible.

These next chapters will contain information and knowledge to highlight that the Bible is an amazing book with God's fingerprints all over it. I will show that any of the arguments put forth to discredit or dilute the truths set forth in the Bible are inadequate to invalidate the claim that the Bible being God's Word. The opponents of the Bible cannot see the forest for the trees. In fact, I would take this metaphor one step further and say that they can't see the forest because of a few leaves on the tree obscuring their view of the entire forest.

Much of the material presented in this section resulted from radio programs I listened to during ten years of commuting to and from my consulting job at NASA's technology transfer center at the University of Pittsburgh. Some of the speakers on radio programs were men like D. James Kennedy, James Dobson, John MacArthur, and Charles Stanley. The books they and their guests recommended are the sources for some of my thinking and information I will be presenting. I first gathered this material for the many adult Sunday school classes I taught and later for the God Discovery Banquet I developed and have presented many times. The great interest and encouragement I received from these presentations is the reason I decided to write this book.

All religions are generally based on the idea of a higher power. When it comes down to it, any person, particularly anyone with charisma and the ability to communicate can create a religion. Founders of a religion may be sincere and truly believe that their ideas, feelings, and experiences are true, but when it comes down to it, are they really true? Is there other evidence that backs up their sincerely held beliefs and proclamations of who God is? The whole premise of Section Two was to show scientific evidence pointing to the existence of God. Can there be scientific evidence to back up their claims about God? And if their claims are counter to the ones made concerning the Christian God, can those conflicting claims both be true?

Many religions logically contradict each other, which means they cannot simultaneously be valid. In other words, one cannot say something is white while the other says it is black and them both be true. The *World Christian Encyclopedia* (2001) states there are 19 major world religions subdivided into 270 large religious groups. The main question for these religions is: "Is what we hold to be true really true?" For the Christian faith, the Bible is the principal and fundamental source for the beliefs about God and His relationship with humans and His creation.

If the Christian faith and the Bible are true as its followers claim, then the Bible should possess trademarks and validations of it being unique and the sole source of information on who God is. When comparing it to the books of other religions, the Bible should have traits that set it apart from the other religious books. Allow me now to present information and evidence that point to the Bible being the true Word of God.

God and the Scientist

In many circles today, good science involves not having any notions of God and should be devoid of any religious influence. Francis Schaeffer, an American theologian and philosopher, in his film series titled *How Should We Then Live* in "Episode VI: The Scientific Age," made the startling observation that the key advancements in science actually came about because of and not in spite of Judeo-Christian thought. Committed Christians like Sir Isaac Newton, Kepler, and Galileo were responsible for many of the advancements that are the foundation for the modern era.

I had the privilege once to see one of Newton's books that talked about his theories of motion. I was amazed that intertwined in his scientific writings were his acknowledgements of the God of the Bible. These early scientists knew God had put forth laws and commandments so people could live in an orderly fashion that matched how they were designed to function. Armed with that information and worldview, they concluded there had to be physical laws which creation obeyed. Their quest became to discover those laws, all the while acknowledging God as that law's originator.

D. James Kennedy in the chapter on science in his book *What if Jesus Had Not Been Born* (Kennedy, 1994) listed 30 Bible-believing accomplished scientists who have the honor of being named as the father of a branch of science. Some "fathers" and their areas of expertise listed include bacteriology (Louis Pasteur);

calculus (Sir Isaac Newton); celestial mechanics (Johannes Kepler); chemistry (Robert Boyle); computer science (Charles Babbage); electronics (John Ambrose Fleming); electrodynamics (James Maxwell); electromagnetics (Michael Faraday); genetics (Gregor Mendel); isotope chemistry (William Ramsey); and gas dynamics (Robert Boyle). From this list, you can see that Christian thought had a great influence in the role of scientific awareness and discovery as it is taught today.

The Bible is neither a science book nor is it meant in any way be a scientific book. That being noted, there are some interesting instances where Scripture presents scientific information that would not have been understood by the contemporaries of the original scriptural accounts. One such example is a metaphor used in Genesis 22:17, which describes a promise God made to a man named Abraham.

In this Genesis passage and a few others that follow, God told Abraham that even though he was childless, his descendants would be as numerous as the stars in the sky and the grains of sand on the seashore. Living in those times, this paired metaphor would be a problem when carefully considered and examined because with the naked eye a person can only count between 2,500 and 5,000 stars.

To have a mental image of this paired metaphor, I went and got sand and counted around 2,500 to 5,000 grains of sand. To my amazement, they did not even fill a tablespoon. This would imply that the metaphor in Genesis was poorly stated. The problem, however, is not a poorly constructed metaphor but one that could not be verified due to a lack of technology at the time. With the advent of the Hubble telescope in outer space, the number of stars that have been seen is literally astronomical (pardon the pun). It has even been stated that the number of stars actually rival the grains of sand on the seashores of the world.

This fact could not have been known when the God-inspired writer recorded this incident and statement in Genesis. It would take thousands of years for the technology to catch up and validate this paired metaphor. Keep in mind that the metaphor described how many descendants Abraham would eventually have.

Abraham was the father of the Jewish people and nation that has existed for more than 4,000 years, during which time there would be many such descendants. In addition, in the New Testament book named Romans, it states that anyone who professes Jesus Christ as Savior and Lord is spiritually grafted into the Jewish race, the family of God. The number of people who are descendants of Abraham are indeed quite numerous as promised in Genesis. When you add the numbers of professing Christians throughout all history to all the ethnic Jews who ever lived who are direct descendants of Abraham, you have an uncountable number that rivals the grains of sand and the stars.

Chapter Twelve

Archaeology and the Bible

For decades, scholars with a team to assist them have sought information through archaeological digs to help us understand how people from past cultures lived, thought, and thrived. Ancient writings have motivated and directed many digs, but the information and stories in the Bible account for the greatest amount of archaeological work. Many claims have been made that certain stories in the Bible are myths or legends since no archaeological evidence exists for those specific events.

However, as the years go on and more digs occur, many of those questionable biblical events have proven to be real through recent archaeological discoveries. In fact, more than 25,000 sites and more than 10,000 records of events and people in Bible have been confirmed by archeological finds! (Kennedy, 1995, p. 35). This contrasts with the Book of Mormon that talks about cities in the North American continent at the time of Christ, none of which have been found from any archaeological work (McDowell, 1979, p. 348)

One example of an archaeological find removing previous doubts surrounding a biblical reference was the existence of the people termed the Hittites in the Old Testament. D. James Kennedy in his book *Why I Believe* (page 30) used the example of scholars in the 18th and 19th centuries claiming that no historian from antiquity ever mentioned anything about the Hittites having existed. The Bible mentions them in eight different chapters in the

Old Testament. The German archaeologist, Dr. Hugo Winckler, went to the area where the Bible indicated the Hittites lived and conducted archaeological digs. The archaeological finds were amazing (Kennedy. P. 30), uncovering 40 Hittite cities, including the capital, along with many monuments that described their exploits and various activities. With this new information in the early 20th century, the Hittites were no longer classified as a myth but were an actual empire just as it was described in the Bible.

GENESIS JOSEPH

Recent studies and work shed even more light on the authenticity of biblical accounts, specifically writings in the book of Genesis in the Bible pertaining to Joseph, the favorite son of Jacob, whose jealous brothers sold him into Egyptian slavery. According to the Bible, Joseph served many masters with distinction and came to prominence in Egypt after interpreting Pharaoh's dream that warned of impending doom from a famine after seven years of plenty. Because of Joseph's skill advising and managing affairs, he was appointed to be second in command to prepare Egypt for the famine, enabling Egypt to live and prosper through the famine.

To escape the famine, his father eventually moved from his homeland in Canaan to Egypt with the sons who had sold Joseph. At the beginning of their stay in Egypt, they enjoyed an elite lifestyle until Joseph died and their descendants eventually becoming slaves for future pharaohs who did not know Joseph. Since this story is not found in any other historical writings from antiquity, the whole biblical story of Jacob, Joseph, and the other sons of Jacob coming and living in Egypt has been seen as a myth. Just like the Hittite story mentioned above, however, archaeological evidence is coming to light that is changing this assumption.

An interesting and well-done film series titled *Patterns of Evidence: Exodus* (Mahony, 2014) provides key evidence from Egyptian archaeological studies that strongly supports the biblical accounts of Joseph, Jacob, and his other sons coming to Egypt. I recommend viewing this series to see all the evidence it presents but let me highlight just a few of the findings from the featured interviews with Timothy Mahony and two Egyptologists, David Rohl and Charles Aling.

Near the delta of the Nile River where it empties into the Mediterranean Sea is the land of Rameses (named after one of the pharaohs) under which archaeologists discovered the city of Avaris. Near the city is an ancient man-made canal still used today called the Waterway of Joseph. The dig uncovered buildings and structures that clearly indicate Joseph settled there along with his father Jacob and brothers. They found in the middle of the city a bigger building with architecture similar to that found in Jacob's native land of Canaan.

Eventually, this structure was enlarged with all the characteristics of an important nobleman owner. The building had 12 columns and 12 tombs adjacent to it, which could be representative of the 12 sons of Jacob. In addition, one of the tombs was pyramidal in shape, a type of tomb usually reserved for royalty. The pyramidal tomb had a large statue twice the size of a person, the size an indication the person buried there was important to the pharaoh and the kingdom of Egypt. The statue had red hair, a mushroom hair style, and a yellow complexion, all traits common to northern Semites from the Canaan area. This would appear to be the tomb of Joseph.

Bolstering the claim of this being Joseph's tomb, no bodily remains were found, unlike other Egyptian tombs. Since grave robbers do not steal bones in tombs because they are worthless, there must be another explanation for the tomb being vacant. Again, the Bible provides a clear explanation. Before Joseph died, he gave instructions that his people were to take his body back to his homeland of Canaan when they left Egypt. Four hundred years later, at the start of the exodus of all Jacob's descendant back to Canaan, the Bible mentions that they did indeed take Joseph's remains with them to bury in Canaan. This is only one of many archaeological finds that seem to have caught up to what the Bible has reported all along.

JERICHO

Jericho is the site of another familiar Old Testament account that has been the focus for archaeologists. Jericho is located in the West Bank, a little west of the Jordan River and north of the Dead Sea. The Israelites made this their first conquest as they entered the Promised Land after leaving Egypt around 1400 B.C. Joshua, an Old Testament book in chapters two through six, described the event when they crossed the Jordon to surround and conquer Jericho when "the walls came tumbling down" as an old children's song states.

After crossing the Jordan River and making camp, the Lord instructed the Israelites to walk around the city carrying the ark of the covenant. They did this for six straight days. They did the same seven times on the seventh day, then blew trumpets and shouted, causing the city walls to fall down thus allowing them to enter and destroy it and its inhabitants—except Rahab and her family. Rahab was a prostitute who helped two spies from the camp of Israel after they snuck into the city to spy out its details. Rahab hid the spies on her roof and finally helped them down over the wall. The spies promised Rahab and her family they would be spared.

There have been numerous archaeological digs at Jericho. Did they make any archaeological discoveries to corroborate this story? First, let's look at the matter of the city wall. An archaeologist named Garstang stated, "As to the main fact, then, there remains no doubt: the walls fell outwards so completely, the attackers would be able to clamber up and over the ruins of the city." In usual battles against walled cities, breaching walls would result in walls falling inward, not outward, thus lending credence to a supernatural event of the outward collapse of the Jericho city wall (Kennedy, pp 29-30). Along the wall they found evidence of houses built next to the wall, which agrees with the Bible stating that Rahab's house was adjacent to the wall, making it easy to lower the spies down outside the wall from her house.

Also, a short section of the wall was found not to have collapsed. Could this be where Rahab's house was located? Another archaeologist named Kenyon was quoted to say, "The destruction was complete. Walls and floors were blackened or reddened by fire, and every room was filled with fallen bricks, timbers and household utensils; in most rooms the fallen debris was heavily burnt, but the collapse of the walls of the eastern rooms seem to have taken place before they were affected by the fire."

In addition, Garstang and Kenyon found jars filled with grain among the burnt debris. This matches the biblical account in two ways. First, the Israelites were forbidden to take anything, except for gold and silver for use in its worship tabernacle. Second, the jars full of grain indicate that the city was defeated quickly with no extended siege, fitting the biblical accounts.

Before I leave this short chapter on archeological evidence that supports biblical accounts, I want to provide one example from the New Testament (McDowell, 1979, p. 112). This specific evidence has to do with the existence of the town of Nazareth where Jesus grew up and lived until His ministry began. The controversy involves the fact that Nazareth is not mentioned in lists of towns in the book of Joshua in Old Testament, in the Jewish historian Josephus' list of 45 towns and villages, or in the Talmud's list of 66 towns and villages. Again, it took some archeological work, this time done by Michael Avi-Yonah, in 1962 in a dig in Caesarea where he discovered fragments that had inscriptions of Nazareth written on them. Josh McDowell lists numerous other findings that support writings in the New Testament as accurate.

In closing this archaeological section, it is fitting to note a quote from a renowned archaeologist, reformed Jewish scholar named Nelson Glueck, who said, "It is worth emphasizing that in all this work no archaeological discovery has ever controverted a single, properly understood biblical statement" (McDowell, 1979, p. 93). I have addressed the historical veracity of the biblical texts. In the next chapter, let's examine the integrity of the texts themselves.

Chapter Thirteen

Textual Integrity of the Bible

Now let's examine the concept of the textual integrity of the Bible itself. Before we do that, let's look at some basic information concerning the Bible. It is divided into two parts, the Old Testament, which are the much older writings before the coming of Jesus, and the New Testament that begins with the birth of Jesus and then describes his ministry and the development of the early church. There are 39 Old Testament books and 27 New Testament books. Although there are different types of literature within the Bible such as history, poetry, books of wisdom, prophecy, laws, contract law, and apocalyptic or futuristic books, they all complement each other and put forth a common theme and message. This has led to a curious theory by some that it would appear a single person wrote the entire Bible.

In actuality, more than 40 human authors wrote the Bible from a wide variety of backgrounds that included a priest, farmer, kings, shepherd, physician, fishermen, tax collector, religious scholar turned tent maker, among others. To complicate matters, the human authors lived at different times spanning over 1,500 years. By comparison, the Book of Mormon and Koran each have only one author. Yet the Bible has a consistency that reads like a single-author work. The only way to explain this, even though there were so many authors over fifteen centuries, is that they were inspired by a single entity—that single entity being God.

Textual Integrity

Textual integrity means the author's original words and information in the texts have not been altered. If there are any alterations of ancient texts, it could be due to mistakes made from scribes who made handwritten copies of the originals from which more copies were made. The potential problem with this can be found in the modern-day "telephone game" played at parties. People line up and the first person whispers a phrase into the ear of the second person. The second person whispers the phrase they heard and/or remembered into the third person's ear, continuing on to the last person in line. This game generates a lot of laughs when the last person recites what they heard whispered and it's compared to what the first person actually said. The difference can be quite different and comical. That's how it could be with one text being copied from another which is copied to make another.

In addition to the text's integrity, it is important that the subject matter of the text be true and factual. We can find a good example of this from American history. Paul Revere is often portrayed as a brave Revolutionary War hero who risked his life riding to Lexington and Concord to warn the colonials that the British troops were coming. There are accounts of this event such as Henry Wadsworth Longfellow's poem titled *Paul Revere's Ride* written in 1860 to commemorate the famous ride. From these stories, Paul Revere was depicted as a true patriot and hero.

My wife and I visited Boston and joined a tour guide in the dress of the time who guided us through historic parts of Boston, sharing the history our country's start. When it came to Paul Revere and his ride, the guide told us the rest of the story. The British captured Revere that night and questioned him concerning the rebels. He reportedly gave up information concerning the rebel cause, though much of the information was common knowledge. Even so, his actions were deemed as treasonous, and he was tried by the colonials but later released. This is an example of how historical accounts can take on fictional, mythical aspects that mar the original facts. In this case, Paul Revere, though an important heroic figure in America's history, had some black marks that have been overlooked or not brought to light so as not to tarnish his heroic image. Did that happen to the stories in the Bible?

Let me make two comments concerning textual integrity using the above example of a historical rewrite in relation to the Bible. First, like the Paul Revere story, there are many Bible stories that have embarrassing moments of failures, weaknesses, and cowardly acts of people. Unlike the Paul Revere story revision that makes Revere look like the perfect hero, the Bible puts forth every aspect of certain stories no matter how embarrassing they are for the characters involved. A few examples would be Noah's drunken acts after the flood, Abraham's fear-induced lies to two different rulers, Peter's denial of Christ three times, and strife in the early church including those seeking elevated positions and prestige. The Bible includes all the awkward and humiliating moments. Second, let's look at the Scriptures and find evidence that corroborates and prove the integrity of the biblical texts.

One argument against the Bible being reliable is that the original written text has been changed and rewritten by later authors and scribes who were recopying the texts. Remember, primitive printing press (invented by Gutenburg around 1440 in Germany) were not available in ancient times and copies could only be made through tedious handmade copies produced by scribes. One important find that counters this errancy-in-Scripture argument comes from a significant archaeological find.

This find, interestingly enough, was discovered quite by accident. The story goes that in 1947, a Bedouin shepherd boy was watching his family's herd of sheep in an area of the Judean desert just north of the Dead Sea and west of the Jordan River. As is common to many boys bored with an assigned task, he engaged in an extracurricular activity to stay entertained. He began throwing rocks into the entrances of small caves in the vicinity where he was. One of the rocks hit and broke what sounded to him like pottery. Upon further investigation, the boy discovered the cave contained numerous pottery jars containing ancient scrolls.

A local dealer of antiques purchased all the jars containing the scrolls. This led other locals to search other caves in the area and many more jars with scrolls were found. The scrolls were written on both leather and papyrus materials. The desert dry air preserved the scrolls, though old and quite delicate, so they were still legible. The caves containing the scrolls were near the ancient ruins of the Qumran settlement believed to have been established around 200 BC by a Jewish sect called the Essenes. The Essenes were a group of ultra-orthodox Jews disillusioned with religious life in Judea and Jerusalem, believing that the leaders had corrupted the Jewish faith by not strictly adhering to the Scriptures. To counteract this deterioration, they moved as a community to a remote place called Qumran in the Judean desert.

Ray Vander Laan in his shot-on-location video series titled "That the World May Know" (2010) includes a portion that describes the archaeological dig at Qumran. From this site, it is clear the community's life was dedicated to studying, preserving, and making genuine copies of the ancient Old Testament biblical text. The scrolls were determined to have been written around 150 BC. After discovering the scrolls in 1947, it took scholars many years to piece together all the scrolls and fragments of scrolls that were written in Hebrew. The amazing part of this archaeological find is that collectively, all the discovered scrolls are writings of the complete Old Testament books, with the lone exception of the book of Esther. When the Dead Sea Scrolls are compared to the Old Testament we have today in the Bible, they are for the most part identical. Thus, this is one of the proofs of the Old Testament's textual integrity.

The New Testament of the Bible also has its critics as to it textual integrity and the accuracy and veracity of the stories presented. One way to validate an ancient text is to find other ancient documents written around the same time that corroborate the information in the ancient text of interest. In his book, Geisler (2013, p. 222) presents a list of other non-biblical works that were written within 150 years of Jesus' earthly ministry which mention events recorded in the Bible regarding Jesus' life. He organized this information in tabular form with the headings at the top referring to specific aspects about Jesus stated in the Bible. This Table 3 is presented on the following page. In this table, the specific events, characteristics, and details provided about Jesus in

TABLE 3

Non Christian Sources within 150 Years of Jesus

Source	AD	Existed	Virtuous	Worship	Disciples	Teacher	Crucified	Empty Tomb	Disciples Belief in Resurrection	Spread of Christianity	Persecution of Christians
Tacitus	111	x			x		x	x*		x	x
Suetonius	117-138	x		x	x			x*		x	x
Josephus	90-95	x	x	x	x	x	x	x	x	x	
Thallus	52	x						x*			
Pliny	112	x		x	x	x		x*		x	x
Trajan	112	x*		x	x					x	x
Hadrian	117-138	x*			x					x	x
Talmud	70-200	x					x				x
Toledoth Jesu	5th Century	x							x		
Lucian	2nd Century	x		x	x	x	x				x
Mara Bar Scrapion	1st - 3rd Century	x	x	x		x	x	x*			
Phlegon	80?	x					x	x	x		

* Implied

Table 3: Non-Christian sources within 150 years of Jesus

the Bible are categorized as: Did Jesus exist? (meaning is He at least mentioned in the other cited text); was he virtuous? was he worshiped? did he have disciples? was he a teacher? was he crucified? was his tomb where he was buried found to be empty? did the disciples actually believe Jesus was raised from the dead? how did Christianity spread? and finally, were Christians in the newly formed church persecuted? All these events are clearly presented in the New Testament.

Giesler (2013) lists 12 different writings of that era that mention at least two of these events. Josephus, a famous and well-respected Jewish historian of that time, recorded nine Jesus' events in the table. Overall, there are 60 examples of corroborations with specific details concerning Jesus in non-Christian writings of that time period. Geisler's work (2013) therefore provides strong evidence of the New Testament's textual integrity.

Let me present one more study that gives additional credence to the textual integrity of the Bible, specifically the New Testament. McDowell (1979, p. 45) examines many documents and writings mainly from the Greek and Roman cultures of roughly 1900 to 2500 years ago, texts that would have been contemporary to that of the Bible's New Testament. In Table 4 is a portion of the information McDowell had in his book. His original table has 16 different ancient documents listed, which I have abbreviated to only four. The last three columns are of particular interest. We do not have the originally written copies of any of these ancient texts. Rather they are copies or copy of copies that scribes hand copied.

Work	When Written	Earliest Copy	Time Span	No. of copies
Homer (Iliad)	900 B.C.	400 B.C	500 yrs	643
Sophocles	496-406 B.C	A.D. 1000	1,400 yrs	193
Aristotle	384-322 B.C.	A.D. 1100	1,400 yrs	49
Caesar (Gallic Wars)	58-50 B.C.	A.D. 900	1000 yrs	10

Contemporary Scholars Have No Problem with the Textual Integrity of these Ancient Literary Works.

Table 4: Textual comparisons of ancient documents

The second column is the date of the earliest copy, which is related to the next column—the time span from the earliest copy found and when it was first written. When looking over the short list I have provided, you can see that Homer's Iliad has the most reliable record of documents listed, which is also true for the McDowell's (1979) original table of 16 texts. The earliest copy we have of the *Iliad* is from 400 B.C. which is 500 years from when it was first written. Five hundred years seems like a rather long timespan until you notice that all the other ancient text time spans are much longer—generally 1,000 years or longer.

The last column, the number of copies we currently have (which are copies of portions of these documents) is also significant for it provides confidence in the level of textual integrity of the writings. Again, Homer's *Iliad* comes out on top with 643 copies. The others are not nearly as respectable with 10 to 193 copies. Even with the large time span of the copies we have after it was written, and with the relatively few copies we have, scholars have no problem with accepting these writings as close to the original text, i.e., having very good textual integrity. Now that we have laid a baseline or standard for textual integrity that utilizes the time span and the number of copies we have today, how does this compare with the books in the New Testament?

Table 5 is the same table as Table 4 except it includes information for the New Testament. Upon looking at and comparing the numbers with the other ancient texts, the New Testament overwhelms all other ancient manuscripts in terms of textual integrity. There is only a 25-year span from a copy of a New Testament text produced after it was originally written. This is almost two orders of magnitude better than the other ancient manuscripts. The same is true when looking at the number of copies of portions of the New Testament we have when compared to the ancient documents. There are an astounding 24,000 copies of portions of the New Testament, compared to the best of the ancient ones, which was the *Iliad* with 643 copies. Therefore, if scholars have no issues in accepting the ancient materials presented in McDowell's (1979) table and the abbreviated one, then there

TABLE 5

Textual Comparisons of Ancient Documents

Work	When Written	Earliest Copy	Time Span	No. of copies
New Testament	A.D. 40-100	A.D. 125	25 yrs	24,000
Homer (Iliad)	900 B.C.	400 B.C	500 yrs	643
Sophocles	496-406 B.C	A.D. 1000	1,400 yrs	193
Aristotle	384-322 B.C.	A.D. 1100	1,400 yrs	49
Caesar (Gallic Wars)	58-50 B.C.	A.D. 900	1000 yrs	10

Table 5. Textual comparisons of ancient documents

should be no issue with accepting New Testament writings as being authentic and having a high level of textual integrity.

ARE BIBLE STORIES MYTHS?

Another area where the Bible is often challenged where its critics claim it contains a large number of myths is in the area of recorded miracles and supernatural events. To explain the miracles and events, these stories are often interpreted as metaphorical since they supposedly contradict scientific laws. The argument often put forth is that science could not possibly support such supernatural occurrences such as the worldwide flood concerning Noah in the Old Testament book of Genesis; the parting of the Red Sea as the nation of Israel made their exodus from Egypt; the survival of three Hebrew men after being thrown into a fiery furnace by the Babylonian king named Nebuchadnezzar; Jesus being born of a virgin; Jesus feeding thousands with just a few loaves and fishes that resulted in baskets filled with leftovers; Jesus raising Lazarus back to life after he had been dead for days; Jesus restoring sight to a man blind from birth; and Jesus himself being raised from the dead after being tortured and then dying on a cross.

If you cannot bring yourself to believe these and other miraculous stories actually occurred and you deny or cannot believe in supernatural events or miracles occurring, I ask that you consider the argument put forth in the science section above. In that section a detailed, science-based discourse was presented as to the impossibility of the life of anything on this earth to have spontaneously occurred by a chance evolutionary process. Instead, it would require a spectacular, out of this world, and supernatural process or miracle outside of the realms of scientific law needed for creation as we know it to take place. Something like that obviously occurred since we and all other living things exist. To state otherwise is to deny science and put in its place, for one's convenience or comfort, a level of faith in science which is quite extraordinary.

But if you will agree that the act by which all living things came into being was by the creation of an all-powerful God, then for all those who find it difficult to believe the supernatural events mentioned in the Bible, such as the short list given above, should now be more ready to accept them as factual. For if God brought into being all of creation, then it should be easier to accept that God could also perform all those miracles.

An even larger miracle occurred as a result of Jesus's death on the cross, something even the learned people of Jesus's day did not understand. A story recorded in Mark's gospel the second chapter highlights this miracle. It concerns a paralytic man who friends brought to Jesus to heal. With scholars of the Hebrew law present, Jesus told the man his sins were forgiven. The scholars and experts were aghast at such a statement, for only God could forgive sins in the way Jesus was doing.

After that, to show He was God who could forgive sins, He commanded the man to get up, pick up his stretcher, and walk home—which he did. This proved that Jesus acknowledged His divine nature. Since the man's sins were unforgiven (as evidenced by Jesus forgiving them), then it indicated that the man was in danger of eternal separation from God. So as significant as the man's physical miracle was, the greater miracle was the healing of his soul. The actual supernatural part of this is that the event on the cross applied to all of the human race for all time before and after the actual crucifixion of Jesus. A more detailed discussion of this will be saved for later in this book.

In the next chapter, let's examine the accuracy of various prophesies or predictions the Bible made over the centuries to see if those predictions did indeed come to fruition.

Chapter Fourteen

Prophecies in the Bible

Many biblical prophecies have been found to have come true many years after they were written, which is one of the most amazing and convincing arguments that God authored the Bible through the inspiration of its writers. Only God can actually foretell the future. This is the one remarkable fact about the Bible of which almost no one is aware, except those Bible believers who regularly read and study it.

How many prophecies in the Bible, specifically in the Old Testament, are we talking about? According to Kennedy (2005, p. 25), there are more than 2,000 prophecies with specific details that were predicted to occur which have come true. For example, he listed a few cities that had specific prophecies proclaimed in the Old Testament concerning their fate. Kennedy (2005) highlighted the cities of Tyre, Samaria, Babylon, and others in the region of Edom as ones specifically stated in Old Testament books of Isaiah, Jeremiah, Ezekiel, and Micah to be destroyed and not to be rebuilt. In addition, the authors provided specific details of how they were to be destroyed and specifics of their end fate.

Tyre is one of the more interesting cases. In Ezekiel chapter 26, in addition to prophesying Tyre's total destruction, never to be rebuilt, the writer also stated that the rocks and timbers from the city would be thrown into the sea and the surface of the city scraped down to the bare rocks such that fishing nets would be laid on top of the bare rock to dry. King Nebuchadnezzar was the first king to begin the destruction of Tyre around 586 BC, when after a 13-year siege he captured and destroyed the mainland city but was not able to touch the island part of the city protected by the Mediterranean Sea to which the inhabitants escaped. The people of Tyre were master seamen, while Nebuchadnezzar and his army were landlubbers. So how in the world could the ruble of the demolished mainland city ever end up in the sea?

The answer is found in none other than Alexander the Great, the eccentric leader of the Greek Empire. He easily conquered all of the coastal cities and towns, forcing them all to submit to his rule—until he came to Tyre. They would not surrender to him, feeling certain, just like in Nebuchadnezzar's case, that the island could protect them. However, they learned of Alexander's tenacity when he ordered all the ruble from the mainland city to be gathered and thrown into the sea, using it to build a causeway to the island. Its completion resulted in the final destruction of Tyre, after which it was never restored or rebuilt. The bare rock left on the mainland after the construction of the causeway is now used by local villages to dry their nets. This is an amazing prophecy that was fulfilled down to its smallest detail. (Kennedy, PP. 18-20)

Let's now look at a set of these prophecies directed toward the city of Babylon. *The New Bible Dictionary* (1962, pages 111-112) provides some basic information on Babylon. The city was established initially as a small town around 2300 BC, built along the Euphrates River in present day Iraq. It became the capital of the Babylonian Empire in which probably its most famous king, Nebuchadnezzar, reigned from 626 to 539 BC. It was one of the largest cities ever built at that time, having a population estimated at around 200,000. The Hanging Gardens of Babylon was considered to be one of the seven wonders of the ancient world.

The Bible states that God used Babylon to punish the nation of Israel for their worship of false gods that caused them to stray from their loyalty to Jehovah. They defeated and captured the nation of Israel and then, as was their custom, deported the captive citizens back to their country to serve and be assimilated into the

Babylonian culture. Needless to say, this was the most influential and most powerful empire of its time.

So, what are the prophecies in relation to Babylon? In the Old Testament book Isaiah 13:19-20 (written around 700 B.C.), it states, "Babylon, the jewel of kingdoms, the pride and glory of the Babylonians, will be overthrown by God like Sodom and Gomorrah. *She will never be inhabited or lived in through all generations"* (emphasis added by the author). Other parallel passages that predicted the same outcome for Babylon are found in Jeremiah 51:12, 29, 37. The key point of this prophecy is that the city would not be rebuilt again.

The Persians destroyed the city around 539 B.C. People attempted to rebuild Babylon a few times, but the efforts were never successful. There were two notable attempts to rebuild, one by Alexander the Great. I can understand an eccentric, egotistical conqueror like Alexander the Great desiring to rebuild the once glorious Babylon. Alexander planned to rebuild Babylon to the same level of its previous magnificence, worthy only of an opulent eccentric ruler such as himself. Uncovered records indicate 600,000 rations were set aside for the soldiers who were to rebuild Babylon (Kennedy p. 25). At this point, it seemed like the prophecy of Babylon not being rebuilt was not going to come true until an unlikely event happened.

Alexander the Great died in 323 BC at the young age of 32. The group of leaders who succeeded him did not have the same passion for the Babylon rebuilding project, and thus they abandoned his plans. As it so happens, Alexander the Great was not the only eccentric national leader with a desire to rebuild Babylon. A modern leader named Saddam Hussein, the dictator of Iraq who was deposed during a 2003 war and later executed, also had ideas of restoring Babylon. Figure 8 shows the ruins of Babylon. Saddam's summer palace, located in the background upper right, was his first step toward rebuilding Babylon.

From the small sampling of Old Testament prophecies just presented, you should have a good idea why it is so exciting to see details that were predicted hundreds of years before that came true in amazing ways. There is a whole other type of prophecy

Figure 8. Contemporary photo of the ruins of city of Babylon

that is even more remarkable, which I will now present, and they are referred to as the Messianic prophecies concerning the coming of Israel's promised King and Messiah.

Throughout the entire period when the Old Testament books were written or compiled, prophets were writing through the inspiration of God of a coming Messiah, a unique individual, called the Son of Man in Daniel's book of prophecy. This man would become the promised leader of Israel after the fashion of their revered King David. The prophecies gave specific details concerning this Messiah that would come true some time in the future of the nation, causing them to be on constant "Messiah watch." The interesting nature of these written prophecies is that the common interpretation of how they would manifest was often not how they eventually unfolded. The people of Israel were expecting more of a military type of a deliverer to save them from their servitude to the nations around them.

But after Jesus' three-year ministry and shortly after his death by crucifixion, all the prophetic details written hundreds and even thousands of years prior to Jesus' death were then seen to clearly pertain to Jesus' life and ministry. To illustrate this, let's look at a few prophecies in detail. The first one is from Isaiah 53:9, written 700 years before Jesus was born. This passage states, "He was *assigned a grave with the wicked, and with the rich in his death,* though he had *done no violence, nor was any deceit in his mouth*" (emphasis added).

In reference to the phrase "nor was any deceit in his mouth" in the above verse, Jesus was above reproach and non-political/military, the opposite of what the Jews had expected because He did not promote or participate in violence. The one exception was when the money changers and merchants selling animals for use in sacrifices temple were taking advantage of the worshipers, completely ignoring the holiness of the venue. Scripture tells of Jesus turning over the money changers' tables and driving them out of the temple area. Jesus' disciples interpreted this later as being (see John 2:17) a fulfillment of the prophecy in Psalm 69:6 where its states the "zeal for your house will consume you." My next focus for this passage will be on two other points in the verse, both of which relate to how his death and burial were to occur.

Jesus was executed between two convicted thieves, and thus His execution and grave were among the so-called wicked thieves as predicted in Isaiah 53:9. How then could Jesus's death also be associated with the rich? Jesus owned virtually no property other than the clothes on His back. This prophecy of association with the "rich in his death" was in fact fulfilled by where he was buried, in a tomb carved out of solid rock that had a large stone that was rolled over the entrance to seal the dead body inside. A tomb of this type could literally take months to carve out using the primitive hand tools of the time. Therefore, this tomb could only be afforded to the extraordinarily rich of the time and thus the prophecy "rich in his death" was fulfilled.

The next Old Testament passage containing four different messianic prophecies, specifically relating to details surrounding Jesus' death is Psalm 22:16b-18, which states, "They *pierce my hands and my feet. All my bones are on display; people stare and gloat* over me. They *divide my clothes among them and cast lots for my garment*" (emphasis added). The first emphasized phrase is "they pierce my hands and my feet," which is clearly predicting crucifixion as the manner of death for Jesus.

The body of a crucified person was secured to the cross by nails that were hammered through their hands and feet, thus fastening their body to the wooden cross planks. The naked body hanging from the cross would naturally accentuate the bone structure, particularly the ribs, which would probably extend to the point where they could be counted. The authorities allowed passers-by and spectators to witness these public executions to discourage criminal activity or rebellion, and those onlookers would cruelly jeer and ridicule those suffering and hanging on the cross as they slowly died. This practice fulfilled the words that "people stare and gloat over me."

Finally, the Roman soldiers would generally take possession of the clothing from those who died on the cross. In Jesus' case, we are told He wore a garment that was seamless and therefore could not be easily divided among the solders. The soldiers witnessed many of these executions and it was not beyond them to callously gamble for this single item of clothing such as the one Jesus owned. When they did so, it fulfilled the prediction that someone would "divide my clothes among them and cast lots for my garment."

Reading the passage above, one would come to the conclusion that it clearly described what happened at a crucifixion, specifically the one Jesus endured. King David wrote this particular prophecy in Psalm 69 1,000 years before Christ was born. What is even more remarkable is that Persians did not invent the use of a cross for execution until 400 to 300 BC, 600 to 700 years after the passage in the psalm was written, which means the prophecy painted an accurate depiction of a crucifixion centuries before it was ever devised.

The passages above from Isaiah and Psalms provide six specific details about Jesus, and in particular about events surrounding His death. I will not go into a detailed discussion concerning other Messianic prophecies but have listed in the Appendix more Messianic prophecies taken directly from a specific website included in that appendix. When you access that site, what you will find are three columns presenting the specific prophesied detail about Jesus. The middle column is the Old Testament passage containing that prophecy while the third column presents where in the New Testament the prophecy is shown to be fulfilled. There are 67 Messianic prophecies listed in this appendix table, which is far short of the total of 314 Messianic prophecies that are listed in Old Testament. This is definitely one of the hallmarks of God's authorship through inspired writers of the Bible.

Chapter Fifteen

What About Other Religions?

When having a discussion about the existence or worship of God, the question is often raised as to the validity and existence of other religions. Throughout my life, I have had the privilege of getting to know and become good friends with people from other countries who were committed believers from other religions. Without exception, they were some of the most moral, caring, and wonderful people I have ever met. Therefore, my aim in the discussion to follow is in no way to disparage them or belittle their faith.

My goal is to do a cursory overview of the main religions and show how they differ from the Christian faith. From a logical perspective, I hope to show you some of their beliefs that are contrary to foundational Christian doctrine, while also noting key similarities between these non-Christian beliefs that counter those found in Christianity.

CHRISTIANITY AND OTHER FAITH TRADITIONS: THE DIFFERENCE

To avoid going into an in-depth treatise on the various religions, which would be the topic of another book, I will refer to an online video that I have used in the *God Discovery Banquet* presentations I sponsor. This three-minute video uses a fictional story utilizing powerful metaphors of the different religions that clearly show the key difference between Christianity and the other faiths. The video is a production presented by Bluefish TV and is titled "A Man in a Hole." I highly recommend searching the Internet to view it even now before you proceed with this chapter. I have seen this video more than 50 times and every time I do, I have a strong emotional response. Let me briefly explain the video if you can't watch it at this time.

The scene at the beginning of the video involves a man who has fallen into a deep hole from which he cannot climb out. The man in the hole gets the attention of various individuals who pass by the hole and they all offer their solutions for how he can get out of the hole. A single narrator explains all the characters, their situation in life, and their communication with each other. The hole represents our sin nature that has caused our separation of God. What answers do the passersby have for this man's sin problem?

From their dress and the narrator's explanation of what they say to the man, it is obvious that each person at the top of the hole looking down at the man metaphorically represents the main religions and philosophies of the world. The first man dressed in an orange robe tells the man to meditate so he can purify his mind and when he reaches nirvana, all his suffering will cease. The man took his advice and mediated in a seated, cross-legged position in a yoga-type pose. After doing this, the man still found himself in the hole.

The next person to pass by the hole was in bare feet wearing a white, loosely fitting robe. He explained to the man that the hole was simply an illusion and did not exist. In fact, the man himself did not exist. The problem was after that person left the man found himself still stuck in the hole—fictitious or not.

The next person to arrive wore a dark robe and told the man to do good deeds and even though he would remain in the hole, the good deeds done may result in him being reincarnated to a better life. Again, it was no help for the man remained in the hole. The next person who looked down also wore a dark robe but this one had a white head covering. He told the man in the hole to pray five times a day facing the east and to perform five tenets. He told him that if he did this the Divine may set him free. The end result was the same: The man remained in the hole.

The last visitor at the top of the hole was in a white robe and arrived at a time when the man was growing weak and weary. The man in the hole could tell right away that this visitor was different than the others. The person at the top of the hole asked the man if he wanted out of the hole and the man in the hole said he did. Immediately, the man threw down a rope into the hole to enable his escape. Then the man at the top of the hole climbed down the rope. He put the man in the hole, who was very weak by this time, over his shoulder and carried him out of the hole while climbing the rope.

This last man was a metaphor for Jesus since upon closer examination viewers can see blood stains on his hands representing the scars from the nails used to fasten him to the cross. Also, the final scene shows the other end of the rope he used was actually tied to a wooden cross in the ground. After all that, the man was finally rescued from the hole.

Let's look at the salient points from this video. Buddhism, Hinduism, and Islam are the religions clearly symbolized by the different people coming to the hole. A common theme in their advice was that the man needed to perform some sort of action or deed to get out of the hole. However, there was no suggested act he could perform that would actually help him out of the hole. All their advice was based on the assumption that the man's only hope was to get himself out. By the way, the fourth person represented New Age philosophy that there was not even a problem and if the man would accept this as fact, his dilemma would be solved.

The two main points driven home to me of this Jesus-figure experience are: 1) the man was unable to get out of the hole by himself; and 2) only the loving and sacrificial act of the Jesus figure descending, gently taking the man, and lifting him out of the hole with no conditions or stipulations was sufficient to save the man. That is the major difference between Christianity and all other religions. All religions attempt to show people how to reach out to God while Christianity reveals God through Christ reaching out to people.

The point is that God is reaching out to people without them fulfilling any religious rules, rites, or ceremonies to get God's attention or to find His favor. This is truly amazing in light of what was presented earlier in this book in regard to God being the Creator of all life. I described the vast complexity and overwhelming number of species which establishes Him far above anything we experience in terms of creative abilities, power, might, or knowledge. The video depicts the extent of God's love as He humbled Himself to become a man, making the depiction of Him climbing down the hole even more remarkable.

I volunteered at a church camp during my summers while in college and graduate school and the camp held nightly outdoor chapels. Each family group was responsible to lead the service for one of the nights. The counselors would plan the service and try to use creative ways to drive home a specific spiritual point. I wrote a skit to help the campers appreciate the extreme transition Jesus made when He went from His all-powerful and majestic position in His heavenly dwelling to come to earth in the flesh, then experiencing the pain and limitations of being human.

The plot of the skit was centered on a construction site. A bulldozer, made from a cardboard cutout, appeared on the scene to move dirt. In the dirt was an ant hill with many ants. A person went over and shouted to the ants that they had to move or they would die. Of course, they did not understand. The person thought about how he could get the ants to listen. He decided to become an ant and try to warn them. In the process of warning them, he got between them and the bulldozer and died. On the cardboard bulldozer we made, we wrote the word *sin*. The main point of the skit was to demonstrate how extreme the process was of God becoming a man, done because of His great love for people. In reality, this metaphor of a man becoming an ant falls far short of the actual incarnation of God into a man. In the next chapter, let me continue a discussion of God's goodness as represented by the cross of Christ and why it is so important that He come down to help us as depicted in the video.

Chapter Sixteen

The Journey

Each one of us takes a different and unique life path. The Bible clearly states in many passages that God first loved us, each person individually, before we can love Him. Probably one of the best known and oft quoted passages in the Bible is John 3:16, often seen on signs at sporting events or on a line of clothing. Why is this passage the overwhelming favorite to display in the hope that curious people will look it up?

This verse is one of the few verses that succinctly states the fundamental path to take if a person wants to know God. Almost all Christians know this verse by heart and may have been the first one they memorized. That was the case for me as I grew up in my family's country church. The verse states, "For God so loved the world that he gave his one and only Son, that who shall ever believe in Him will have eternal life." Let's look at this verse in more detail. The verse says, "God so loved the world" but the word *world* does not refer to the physical world but rather to the people in the world. It is clear that knowing God begins with His love and not from any attempts we can make to reach or connect with Him. This truth and realty are found again and again throughout Scripture.

Next, it is clear that God himself, incarnate as His Son, made that transition from heaven by coming to earth and giving His life on the cross. The idea "gave His life" means He willingly died on the cross and is found in other Bible passages to reiterate the fact that He did indeed die. The verse also presents the idea of believing in the Son, Jesus, which means having faith that He is God, trusting what He did by dying on the cross for each person. It is also accepting that He miraculously came to life again after three days, demonstrating His mission was accomplished. Then He was accepted into heaven through what is known as His Ascension. The idea of believing is not simply affirming that Jesus existed in history. Rather it is a life-transforming belief that once accepted releases His grace to apply to one's life and being. This fundamentally changes every aspect of a person, from how they think, what they do, living their lives through the selfless caring for others as Jesus did, and telling others of His love.

There is another key concept that needs to be taken from this verse that may not be obvious at first glance. A question could arise after reading this passage: Why did God have to give His one and only Son? The answer is found in other passages in the Bible. One clear explanation is found in Romans 3:23 which states, "All have sinned and fall short of the glory of God." The sin spoken about here does not necessarily refer to individual sins such as murder, lying, adultery, or covetousness, but rather a condition with which all are born.

It is the condition of being separated from God and having a nature that is focused first and foremost on ourselves and our own perception of who we are and what is best for us. This in general will lead to many of the sins mentioned above but will also lead to many of the more subtle sins such as pride, lust, selfishness, self-serving actions and thoughts, and lifestyles and philosophies contrary to God's will or purpose, which Jesus addressed when He taught on earth and left His teachings for His followers to disseminate.

Jesus said, "I tell you that anyone who is angry with his brother will be subject to judgement . . . but anyone who says, 'You fool!' will be in danger of the fire of hell (Matthew 5:22b). Jesus also said, "I tell you that anyone who looks at a woman lustfully has already committed adultery with her in his heart" (Matthew 5:28). Some may wonder what is so wrong or harmful if they have lustful thoughts of another person or if they have anger toward someone.

First, thoughts can be the first step to actually carrying out an action. Unchecked thoughts may dictate how we treat another individual or relate to them in a self-absorbed as opposed to a loving, selfless way where we put their well-being above ours. This concept of selflessness is beautifully state in the New Testament book of Philippians where Paul wrote that our "attitude should be the same as that of Christ Jesus" (Philippians 2:4). Also, "in humility consider others better than yourselves. Each of you should look not only to your own interests, but also the interest of others" (Philippians 2:5). Harboring thoughts of continual anger or lust to another individual is the opposite of making their wellbeing our highest priority.

When we look at this from God's perspective, who the Bible states that He clearly sees and knows our thoughts and motives, we come away with even a deeper understanding. This truth is clearly stated in Psalm 139:1-4:

> "O Lord you have searched me and you know me. You know when I sit and when I rise, you perceive my thoughts from afar. You discern my going out and my lying down; your are familiar with all my ways. Before a word is on my tongue you know it completely."

Imagine if you possessed for a day this omniscient ability and were able to see everyone's thoughts as clearly as you see their actions. I am sure your opinion may drastically change concerning people you know. What would be even worse if others had that ability and could see your thoughts and motives. This would be another incentive to clean up your thought life, thus making seemingly innocent thoughts a much more serious matter.

Then let's consider God's holiness. By definition, holy means to set apart and above anything else. God is set apart from all as holy and cannot, because of His character, be in the presence of sin without the proper punishment and justified judgment against it. If God is a perfect judge, then any infraction cannot go unpunished or God's perfect justice would not be perfect. This brings us back to the need for God giving His only Son to die on the cross to take the punishment for our sin upon Himself. And that also addresses why unholy thoughts are not a victimless sin. Those thoughts put the one thinking them in an unholy state before God.

Because Jesus was fully God but also fully man, He was the only sinless person to live, making Him the only sacrifice capable of taking on the sin of all people for all time and thus enabling us to be in fellowship with God. This whole idea of simultaneously fulfilling God's inherent characteristics of having perfect justice carried out while at the same time being totally merciful and loving through the actions of Jesus Christ is phenomenal and for many may be hard to comprehend. I once heard a story that truly sheds light and provides a wonderful understanding of this idea of the simultaneous satisfaction of God's justice, mercy and love.

The story goes that there was a judge in a courtroom pronouncing a sentence on a guilty individual who had committed a heinous crime worthy of a severe punishment. The judge pronounced sentence, which included a lengthy jail sentence that all in the room, including the criminal, fully expected and saw as necessary and right. What happened next, however, was totally unexpected and shocking.

The judge came from behind his bench of justice, stepped down to the level of the criminal, removed his judge robe, and put his hands forth in the direction of the bailiff to be handcuffed and taken away to serve the criminal's sentence he had just proclaimed. You see, the judge was also the criminal's father. The father had a deep love for his child and as a result offered to serve the sentence instead of his child.

This story provides a vivid picture of what actually happens when anyone accepts Jesus's death on the cross as He carried our penalty for our sins. Jesus in His sinless form stepped down from His judge's bench in heaven, came down to earth, lived among us, loved us, and showed us mercy. He willingly left behind His power, humbling Himself, and went to the cross to suffer and die for the punishment designated for us.

Therefore, God now sees us as not having sin any longer since Jesus took the punishment for our sins by dying on the cross to satisfy God's inherent "perfect justice" nature. The end result was that God's immutable loving character paid the penalty for our sinful nature which otherwise could not be in God's presence without proper restitution. Those who accept Christ's punishment for their infractions are now seen by God as righteous.

Jesus through His act of love on the cross established a personal direct relationship between God and each individual person without the need of an intermediary. The only known activator, however, is to personally accept this sacrificial gift that is free to all, but was obviously not free to God. In other words, Jesus climbed down the hole and rescued anyone who cried out for help, putting their trust in His ability to save them. Once any individual comprehends the magnitude of God's gift of Jesus dying on the cross, there should be an immediate commitment to serve Him out of gratitude and love.

The events and decisions in a person's life that led to becoming a Christian are each unique, but as stated above, there is a common element: God makes the first move. The Bible mentions this in a number of places. First John 4:19 in the New Testament states, "We love because He first loved us." Titus 3:4 says, "Saved us by His love and mercy, not by our righteousness." Ephesians 2:4 declares, "Because of His great love for us, God, who is rich in mercy, made us alive with Christ even when we were dead in our transgressions."

To better understand how each person's journey of knowing God comes through Him first loving us, think of a picture puzzle. Imagine the picture depicts an illustration of an overall understanding of who God is. Upon the addition of each puzzle piece, one receives a better understanding of God. Imagine seeing the very first piece of the puzzle. It could so overwhelm someone that they want the other pieces God provides to see a clearer picture of God, which then leads to a more vibrant and meaningful relationship with Him.

The order and number of pieces one receives will vary. It may take more than one piece to bring a person to see Christ clearly and become a Christian. For me, the first one I was given and accepted was at church camp at nine years of age. The Bible stories, the songs sung, and times around the campfire thinking about God and His Son Jesus' death on the cross all combined to hit home. I felt His love for me. I accepted that first puzzle piece

He handed to me and thus accepted His free gift of salvation for my sins by His work on the cross. As life went on, I received more pieces from church, Bible studies (both personal and group studies), serving others, prayer, miracles, Christian retreats and conferences, and from friendships with other Christians. Each piece helped provide a more complete picture of God and my relationship with Him and others.

That first puzzle piece was an act of God reaching out to me and may seem subtle and vague to you. It was an act of God's mercy because I did not go to camp to look for God. My parents urged me to go, and I had no idea what to expect, except that I would be sleeping in a cabin with other boys and a counselor—while also having fun swimming, playing games, and spending time in the woods. God reached out to me there and I met him there.

God provides many different puzzle pieces that are clearly His first moves in reaching out to an unbeliever. Another amazing, clearly supernatural act of God taking the first initiative is His use of dreams. Some people who did not have a relationship with Jesus or know anything about Him have had crystal-clear dreams in which Jesus clearly revealed Himself. An article *Visions of Jesus* (Fetherlin, 2013) chronicles such a dream-related event.

The story begins with two Christians living in a distant country who had an urging from God to deliver Bibles across the border into a country where there were very few Christians. Not knowing where they were going, they crossed the border with Bibles in a heavy rainstorm, and along the way got stuck in the mud. Not knowing what to do and while deciding their next step, a person knocked on the car window asking if they had brought the books. This person and his family had been waiting a few days for them.

When asked how they knew they were coming with Bibles, the man said he had multiple dreams of a man in white telling him to wait for a book that would be brought to him. The amazing thing is his two sons had the same dream. This dream and its revelation of Jesus and God reaching out to this man in such a miraculous and personal way was his first puzzle piece to help him understand the true God. God made the first move and showed His love toward this man and his family through God-given dreams. God added many more pieces to the puzzle for this man and his family as they read those Bibles. He and his sons and their families along with many others in the area became Christians. In this case, you can clearly see how God made the first move reaching out to these people.

Jesus appearing in dreams occurs frequently in the Muslim world, for dreams are considered a source of revelation for many Muslims. The result is that some Muslims are giving their lives to Jesus and becoming committed Christians, knowing this decision will likely result in their families and communities ostracizing them and could lead to persecution and even death.

It seems that when Muslims see a Jesus figure in their dreams, it is accompanied by an overwhelming sense of love and peace, something they had not previously experienced. That becomes their first puzzle piece about God that Jesus is the source of love and peace. Also, many see the Bible as a book from God since many heard in the dream specific never-before-heard verses being quoted from the Bible. When they recounted what they heard to Christian pastors, the pastors would show them those words from the dream in the Bible. The dreamers would become even more amazed, fully committing their lives to Jesus and the reading of the Bible.

Another vital puzzle piece of information about God is that He is the creator and sustainer of all things. This theme of God being the creator is stated in many passages throughout the Bible. Perhaps the most significant verse is found in Genesis 1:1: "In the beginning God created the heavens and the earth." The scope of God's creation as stated in this verse is inclusive of everything we see and do not see, represented by the phrase "the heavens and the earth." There is nothing that exists beyond these two realms.

This knowledge of God as creator of all things brings into sharp focus God's claim of ultimate authority, something that people, cultures, and philosophies always seem to challenge. People often question what right God has to tell anyone what to do or not do. "I can decide what is best for me" is often their life philosophy. The problem is that God has created all and therefore He has every right to put forth the rules and standards He expects.

Fortunately, God is a loving, compassionate and merciful God who puts forth these standards for us to follow for our own good. Since He created us, He knows how we are supposed to live in order to have a full life that centers on a relationship with Him. Genesis 1:26 states that He created mankind in His image. Our species is the only one of the millions of species to have this distinction. The human race is the pinnacle of creation, our purpose including the ability to know God and have a personal relationship with Him—unlike any of the animals He created.

The problem in most modern societies is what I call the "the upside-down image syndrome." As stated by the Bible, we are created in God's image. Some turn that upside down and instead make God in their image. They will craft God—who He is, what He stands for, and what He requires—according to what they are comfortable with and not according to who He really is. This tendency even exists within Christian circles. Recently I heard a of a nondenominational group call the "Red Letter Christians." The first four books in the New Testament that deal solely with Jesus's time on earth are called the gospels and the words spoken by Jesus in those gospels are printed in red in some Bible translations. The intent of this group is to focus mainly or exclusively on those words while avoiding some other passages in other parts of the Bible with which they have problems understanding or applying in their lives. Thus, they are creating an image of the God they want to serve.

This position of focusing only on Jesus's words as being the primary authoritative portion of the Bible, thus creating the idea that only certain parts of the Bible should be considered God's Word, contradicts the evidence presented above illustrating all sections of the Bible have the markings of being God's Word. The combined evidence from archeology, textual integrity, science, and prophecy found throughout the Bible and not just the red-letter sections shows the Bible as a whole needs to be accepted as God's Word and not just the sections one feels comfortable with reading or believing.

There are countless other shapes and sizes of puzzle pieces that are God's first pieces. God reaches out to a person and meets them where they are. That is why there are so many diverse circumstances that explain how people become Christians. As an elder in a church, I had the great privilege of hearing the testimonies of people desiring to become a member of our church. Part of the membership process was for an elder to meet with each new member, with part of the meeting devoted to the new member telling the elder how they became a Christian. None of the testimonies given were the same. Some were similar to mine when people encountered God touching them through a church function. Others involved stories of seeing God work through a life crisis. In each one of the stories, however, God reached out to them first.

Another one of the different puzzle pieces is the revelation of hell and heaven from near-death experiences. Over the past few decades, the technology and means to revive someone from a stopped heart have resulted in stories of people experiencing the afterlife, both of the heaven and hell variety. In some cases, this resulted in their accepting the salvation offered by Jesus. The next section will briefly discuss the idea of heaven and hell and near death or death bed experiences.

Chapter Seventeen

Heaven and Hell

The existence of heaven and hell along with who goes there are controversial topics—and that is especially true of hell. In general, everyone wants to believe that all go to a blissful place called heaven after death. Hopefully at this point, your automatic response should be, "I wonder what the Bible says about these topics?" In a nutshell, the Bible says heaven is the place where God reigns, with Jesus seen as God and being worshipped sitting at His right hand (figuratively speaking). It is the place where all who know Him will be in His presence, worshipping continually and joyfully. It is a place where there are no more tears and suffering, and where we will live in a place set aside to carry out tasks assigned to us.

As you would expect, there are a number of Bible passages that talk about heaven. In Matthew 7:21, Jesus was delivering His sermon on the Mount and stated, "Not everyone who says to me, 'Lord, Lord,' will enter the kingdom of heaven, but only he who does the will of my Father who is in heaven." We learn here from Jesus that there is a heaven where God the Father resides and determines who gains entrance.

What the Bible Has to Say

Philippians 3:20-21 states a Christian's "citizenship is in heaven. And we eagerly await a Savior from there, the Lord Jesus Christ, who by the power that enables him to bring everything under his control, will transform our lowly bodies so that they will be like his glorious body." This passage describes how followers of Jesus wait with eager anticipation of being in heaven with a glorious body that has none of the ills and pains of this world. Revelation 21:3-4 provides still more details concerning heaven and states,

> "Now the dwelling of God (heaven) is with men, for the old order of things has passed away, and he will live with them. They will be his people, and God himself will be with them and be their God. He will wipe every tear from their eyes. There will be no more death or mourning or crying or pain, for the old order of things has passed away."

Hell is described in opposite terms. It is the place of eternal torment "where their worm does not die and the fire is not quenched" (Mark 9:48). It is the place where those who have not accepted salvation through Jesus's sacrifice on the cross will go after physical death in this world. They will not experience any of the goodness or bliss that come from being in God's presence in heaven. There is an impenetrable barrier separating the sufferings of hell and the magnificence of heaven.

This separation is mentioned by Jesus in His parable found in Luke 16 about a rich man who ended up in hell while a beggar named Lazarus ended up in heaven. The rich man asked that Lazarus be dispatched from heaven to bring him comfort in hell. The response was "between us and you a great chasm has been fixed, so that those who want to go from here to you cannot, nor can anyone cross over from there to us" (Luke 16:26). So once someone is in in heaven, they will always be in heaven and the same is true for those in hell.

Verses in the Bible that describe hell provide a drastic contrast to the descriptions of heaven. Matthew 25:46 quotes Jesus concerning the eternal destiny of followers and non-followers of Jesus: "Then they will go away to eternal punishment [non-followers], but the righteous to eternal life." Note the key word here is *eternal*, indicating an eternity for all with two very different destinies—punishment the destiny for non-followers of Jesus and eternal life or heaven for His followers.

Unbearable agony and unquenchable fire are common key descriptions of hell. Matthew 13:49b-50 reads, "The angels will come and separate the wicked from the righteous and throw them into the fiery furnace, where there will be weeping and gnashing of teeth." Revelation 21:8 states, "The cowardly, unbelieving, the vile, the murderers, the sexually immoral, those who practice magic arts, the idolaters and all liars – their place will be in the fiery lake of burning sulfur. This is the second death."

NEAR-DEATH EXPERIENCES

During a thirteen-week adult Sunday School class I developed and taught, I devote one entire class to the concepts of heaven and hell. At the end of the thirteen-week class, I hand out a feedback survey and I am surprised that many say the heaven and hell class was the most enlightening. In addition to going over some Bible passages concerning heaven and hell, I talk about near-death, end-of-life, and deathbed experiences that have been written and talked about. Kennedy (1999, pp 77-79) recounts an interview with an individual who had a near-death experience that included an experience in hell. The man interviewed gave him permission to tell his story. This person was an avowed atheist who did not believe in God, the Bible, heaven, or hell. He felt once someone died, that was all there was. Then he had a heart attack but was revived.

Before being revived, he had a horrific out-of-this-world experience. He sank into deep darkness while he tried to push a stone into a pit. The pain he experienced all over his body was unbearable, and nothing he could do would lessen it. Asked if he had ever experienced any similar pain during his life, he told of an experience as a kid when a train ran over his leg, almost severing it. He concluded the pain from that did not come close to the pain he felt in the near-death experience.

Kennedy (1999) then asked if he ever experienced the pain of a burn, which is known to be one of the worst kinds of pain. He said one time in his workshop a can of gasoline spilled onto his good leg and caught fire, rolling up his pant leg to show the burn scars. When asked how the pain in hell compared to the burn, he said the pain during the near-death hell experience was many times worse. Needless to say, after being revived from near death, this gentleman became a committed Christian, which is why he wanted to share his story to help others come to faith in Jesus before it was too late.

There are fewer hell-like near death-experiences recorded than there are for heaven. A few reasons have been given for this. The first is that people are too embarrassed to recount their hell experience. The second is the experience was so terrifying that in order to spare the person the horrible agony of the experience, it is pushed deeply into the subconscious to protect the individual. This type of targeted forgetfulness is a known mechanism of the brain to protect a person from an overwhelmingly painful past experience.

Rawlings (1993) experienced a patient that had a hell near-death experience. As he described in his book, the patient was undergoing a stress test on a treadmill to diagnose possible heart problems when he suddenly collapsed after his heart stopped. What happened next was life-changing for not only the patient but also for Rawlings (1999). As he and the nurse carried out their resuscitation procedures, the patient would temporarily become conscious with a horrified look on his face, shouting not to give up because he did not want to go to hell.

This happened numerous times until the patient asked what to do to get him out of hell. Not knowing what to do and with the nurse looking to him for an answer, Dr. Rawlings made up a prayer on the spot for the man to recite that entailed the patient asking for Jesus to save him. Once the patient recited the prayer, he immediately became relaxed and then was brought back to life. After that, the patient became a committed believer, and this encounter also caused. Rawlings (1999) to come to faith. That was His puzzle piece of God reaching out to both men—and the nurse.

Rawlings (1999) proceeded to put together a collection of studies that contained not only heaven-related near-death experiences but also hell-related ones which became the focus of his books. Out of the many examples he puts forth, the one that I will highlight from *To Hell and Back* is a patient who was a Baptist

Sunday School teacher who had two near-death experiences. The first was a horrific experience due to a heart attack during which he witnessed snakes and horrible fire.

After he was revived, he made a sincere, genuine confession and life-yielding decision to follow Jesus, something he actually knew he had not done even though he had been a faithful church attender. Later in life, he had a second heart attack that included another near-death experience. This time his experience was the complete opposite with a wonderful heaven-like encounter. When he did die later from cancer, it was a peaceful death since he knew his eternal destination.

Aside from near-death experiences, some have written about death-bed experiences. Rawlings (1999) has a section in his book on this topic. Positive deathbed experiences have included feelings of peace and contentment. Some have witnessed angels visiting or a wonderful light and an overall readiness to go onto a better place. One of the more amazing end-of-life stories was that of Charles Darwin, the person who put forth the treatise of evolution.

In his last days of life, he recanted his theory, saying he had been young and uninformed when he formulated it. From his reaction and emotional response after recalling this work, it was clear he was stressed and uncomfortable with the fact that his theory of evolution had become so popular to the point of being a kind of religion. In his last days, he focused on and enjoyed reading the Bible, with the book of Hebrews, a book he termed the *Royal Book*, being one of the last he studied. His last request was to have all the servants, tenants, and locals gather to hear a talk concerning salvation in Christ.

Rawlings (1999) also wrote about unpleasant death-bed experiences. One in particular that was written was in regard to the death of Voltaire who had a stroke after which he lingered for many days before he died. Voltaire was a self-reliant, educated man who coined the well-known phrase "the pen is mightier than the sword." In the midst of great agony during his prolonged death, he blasphemed God and raged against men and friends. He even tried reciting various incantations to no avail. His death was so terrible that his nurse who watched him die vowed she would never again see the death of an infidel for all the money in Europe.

Rawlings (1999) told of others' deathbed experiences involving beings no one else but the person dying could see hovering around them, laughing and waiting for them to die—supposedly to usher them to the underworld. There are fewer of the negative deathbed stories when compared to the positive. Rawlings (1999) surmised that there may be many reasons for this, such as being embarrassed to tell about a family member having such an experience or maybe people believing it was a hallucination and therefore meaningless.

As stated above, positive death-bed experiences are completely opposite of the negative in that there is peace and joy often with visions of angels and wonderful light. Dr. Rawlings bemoaned that in today's society fewer people get to experience their loved ones' deaths due to the person being placed in nursing home facilities to die where they are sometimes sedated. Thus, their families miss the opportunity to see dying loved ones enter into heaven. In the past, families would gather in vigilance near the bedside waiting for their love one to pass on.

Rawlings (1999) relates just such a missed opportunity of a woman at a nursing home who was dying of cancer. She was all alone except for him. The family avoided her because she was seeing unusual things and the unpleasant aroma in her room. In this patient's last moments, Dr. Rawlings reported she was calling for her son. Standing in for the son, he said he was there and held her hand. She then said, "Can you see the chariot and the angels? Do you see them? It's glorious. Be a good boy, Mark. Let others know I have gone." Her life here on earth ended and, as Dr. Rawlings reported, the family missed this amazing event.

Kennedy (1999) stated that from the thousands of deaths recorded in many of the books in his library, there was a clear difference between the death of evangelical Christians and those of others. Even a non-Christian psychiatrist whom Kennedy (1999) knew stated that after witnessing many deaths, he could immediately distinguish an evangelical Christian's death from all others. Some who have had that heaven-like experience speak of traveling toward a light and a peaceful place, seeing friends, family and even Jesus just before being resuscitated and brought back to life. This doctor had not only researched and found this type of near-death experiences. On the contrary, he discovered there were also actual near-death experiences of those experiencing hell.

You may ask how as a scientist I can believe that heaven and hell actually exist. I believe they do since the Bible clearly states they do. Accepting that the Bible is God's Word and relying on the evidence described earlier about the Bible being God's Word, I believe they exist although I do not understand or have all the answers about these places. I will say, however, that there is an interesting concept in physics that may provide an explanation or a clue of where heaven and hell are located.

As humans, we live in a world that is defined by four dimensions. Three dimensions identify a location, usually designated as the x, y and z coordinates. Two of the dimensions, x and y, can be represented by a grid that has two axes. The x axis runs left to right on a grid while the y axis runs perpendicular to the x axis going from bottom to top. A location can be defined on the grid by giving the x and y numbers for that point.

For instance, a location (x and y) can be defined such as (1, 3) which means x=1 and y=3. To locate this on the grid, move from zero horizontally to the right one space, and then move three spaces up vertically. This would be the location of a point defined as (1, 3). To move to a three-dimensional representation for defining a location, a third axis called z needs to be added, which is identified as an axis that moves perpendicular to both x and y axis and can be seen as moving out and below from the graph paper.

The location (x, y, z) of (1, 3, 4) is represented on the grid as moving from zero the same distances as those for two-dimensional example (1, 3) and then add the movement of moving the z defined point four spaces up and out of the paper perpendicular to both the x and y axis. The last dimension we experience is time and is defined as (x, y, z, t), whereas for (1, 3, 4, 30), the same x, y, and z are used in the previous example and so would add the time variable to this set of x, y, and z positions.

The difference is the time which is stated as 30. If the units used are seconds, then that location only applies for the time 30 seconds from the start at zero seconds. Let's say at time 30, my position puts me at my dining room chair in my home but changing the time to 24 hours later will more than likely have a whole different set of positions x, y and z. Where am I going with this idea of dimensions?

There are theories proposed in physics that suggest there are a total of ten dimensions, with string theory even saying there are a total of 26 dimensions (see http://fancyfrindle.com/how-many-dimensions-are-there/). In our limited finite existence on Earth, we are only limited to experiencing four dimensions. Might the heaven and hell dominions described in the Bible, where the two cannot meet or be connected, actually be defined by different dimension variables? Yet God, from His all-powerful position, can transition between all these dimensions which He created.

This could also explain where the spiritual battles described in the Bible's book of Daniel are occurring. They could be happening in another parallel dimension to ours. There are also biblical descriptions of angels ministering to humans to help and guard them. The common phrase used to describe them is "guardian angels." Along the same line is an explanation for how Satan and his demons are also running rampant among us, creating havoc.

A good fictional writing that helps one understand the intricacies of this spiritual battle going on is the three-book series by Frank Peretti with the titles *This Present Darkness, Piercing the Darkness,* and *The Prophet*. These books follow the main character who is a newspaper reporter and include stories of strife and cultural combat, while at the same time describing the activities of angels and demons in a parallel universe where they influence people and their way of thinking and acting.

Thus, physics theories of more dimensions could possibly explain the existence of the spiritual battles that can affect us in our four dimensions. In the Bible, the sudden appearances of individuals, such as Jesus appearing to Paul on the road to Damascus and Jesus' mystical, sudden appearance after the resurrection to the disciples in the locked upper room could also be explained by the supernatural movement of Jesus from one dimension to another.

Before leaving this section, let me make one last point concerning hell. It is something that deeply concerns God, for the Bible states that it is His desire no one goes there after death. The extent of His concern was the drastic measures God took to provide a way to heaven instead of hell. It is a free gift to each of us, but it cost God everything. God became a man, and Jesus took on all the limitations of being a human, including experiencing all the pain and suffering of this world. Jesus still retained being fully God capable of performing miracles and having deep, accurate insights into people. He then willingly died on the cross to take the punishment for our sins to satisfy God's character of perfect justice.

The other information proving God's desire to save people from hell where they will be separated from Him for eternity is His patience and delay in returning to reign and bring final judgment. The Bible states that Jesus will return to defeat evil once and for all, throwing those condemned into hell forever. One passage in the Bible that states this desire of God is the New Testament book of 2 Peter 3:9 where it is written, "The Lord is not slow in keeping his promise, as some understand slowness. He is patient with you, not wanting anyone to perish, but everyone to come to repentance."

Conclusion

Your Journey

I do not know where you are in your spiritual or life journey. You may be a believer, having accepted His gift of salvation and sincerely yielded your life totally to Him. If that's the case, this book has hopefully encouraged you in your faith, providing a few more puzzle pieces to help you obtain a more complete picture of God. Or maybe you are a church goer, like the Baptist Sunday School teacher mentioned above, who knew he really had not made a full commitment to Christ but was instead going through the motions. Perhaps you are someone who thinks just being in church and a regular attender are good enough. The thinking that church goers can become Christians by sitting in church is no more possible than a bicycle becoming a car simply by spending a lot of time in a garage. Church going is not enough without a firm commitment to obey the commands of Christ in all of life. I hope the information in this book will help you take the next step of being more than someone who attends a church, but rather a 24/7/365 believer and follower of the Lord Jesus Christ.

Maybe this book provided answers to some nagging doubts and questions that have prevented you from committing totally to Jesus. These questions could involve doubts that God miraculously created everything as opposed to evolution being the source of life, or maybe doubts about the reliability of the Bible, or possibly the lack of evidence to key points of the faith such as resurrection, miracles, and God's all-knowing, all-powerful, and omnipresent nature. If it took God to bring about all things, living and nonliving in all their complex order, then believing that God can pull off the resurrection, the miracles and have all the God-like characteristics should be much easier to believe after seeing the information presented in this book. Or maybe at this time you have very little knowledge of the Bible and who Jesus is when someone gave you this book. Or perhaps you somehow found this and it piqued your interest—thus being a piece of the puzzle through which God is making the first move toward you to reveal Himself to you.

What must one do to enter into a relationship with God that begins here on earth and continues on into eternity in heaven? After realizing there is a personal and Almighty God who exists, you then need to understand and accept that according to God's standards, you, along with everyone else, have fallen short of being right before God. Roman 3:23 states that "all have sinned and have fallen short of the glory of God."

God is intimately acquainted with you and every other created being. After going through my science section where you learned the knowledge and intelligence needed to put together the proper order of DNA base pairs (I.e., the rope ladder rungs) for the more than seven million species of animal and plants and how the billions of cells function and interact within a living being, it is like child's play for God to know you intimately, including all your thoughts, actions, and motivations.

God's inherent character is that of a being all perfect and powerful, attributes that are clearly seen through His creation and creativity. This perfect character by definition cannot allow anything sinful or contrary to His nature, will, or plan to be in His presence—let alone in a relationship with Him. Isaiah the prophet experienced this firsthand during his supernatural encounter with the living God as recounted in the Old Testament Bible book Isaiah 6:5" "'Woe to me!', I cried. 'I am ruined! For I am a man of unclean lips, and I live among a people of unclean lips, and my eyes have seen the King, the Lord Almighty." Anything we may try to do including any good works fall short of pleasing God. As it says in Isaiah 64:6b, "All our righteous acts are like filthy rags" before Him. Things like pride in your good works, ulterior motives, showing off your good deeds and a hollow humility ultimately attempt to put the spotlight on you and take away the glory due God.

In reality, there are no amount of good works that can earn a relationship with God and a place in heaven after you die. The good works of people are as if every living person would line up on the California coast, jump into the Pacific Ocean, and attempt to swim to Australia nonstop for a total of 8,128 miles (13,080 km). Many would drown within the first mile, others may make it to 5, 10 and maybe even 20 miles. The furthest anyone has ever swum was 140 miles, much shorter than the 8,128 miles.

It is the same with any good works we try to do in order to measure up to God's inherent standards. Sufficient deeds cannot be done. Everyone comes up short just like the swimming challenge above. However, God in His great love has done it all on the cross to pay for your sins, enabling you to come into His presence and have a loving relationship with Him. It would be like Jesus swimming up to you and putting you on His back and then swimming to Australia for you—or climbing down into the hole to carry you out as I described earlier.

It is interesting that at the very time Jesus died on the cross, the curtain to the Holy of Holies in the Temple in Jerusalem that prevented anyone except the designated high priest from approaching God holy place was supernaturally torn down the middle (see Mark 15:38). This indicated that Jesus' death removed any barrier between any individual and God through His death.

Romans 6:23 states, "For the wages of sin is death, but the gift of God is eternal life in Christ Jesus our Lord." The consequences of not measuring up to God's standards (which no one can) and living a self-centered life apart from God are death and eternal separation from Him. The evidence of not living up to God's standards covers a wide spectrum of activities. There are things that a person has done or is in the midst of doing that may seem so terrible that the person feels there is no way God could ever accept them. These sins include murder, spousal or child abuse, drunkenness, bursts of anger, abortions, or addictions, just to name a few. One short list is provided in 1 Corinthians 6: 9-11 which says:

> Don't you know that evil people will not receive God's kingdom? Don't be fooled. Those who commit sexual sins will not receive the kingdom. Neither will those who worship statues of gods or commit adultery. Neither will men who are prostitutes or who commit homosexual acts. Neither will thieves or those who always want more and more. Neither will those who are often drunk or tell lies or cheat. People who live like that will not receive God's kingdom. Some of you used to do those things. But your sins were washed away. You were made holy. You were made right with God. All of that was done in the name of the Lord Jesus Christ and by the Spirit of our God.

Note even though many of the Christians Paul was writing to in Corinth committed the sins in this short list, verse 11 states that they used to do them but were washed clean and made right before God through Jesus Christ. So, no matter what you have done, God will forgive you if you sincerely ask for forgiveness and turn away, with the help of God, from these sins to follow Jesus.

If one is truly sorry and comes to God in repentance (a level of sorrow that results in making, with God's help, a 180-degree turn in the opposite direction in life and renouncing one's sin), then God will save that person and, as it says in 2 Corinthians 5:17, God will make them a new creation with a fresh start. But the doer of good deeds also falls short of God's standards, but their lifestyle may make it more difficult for them to acknowledge their need of a Savior.

As we have seen in Romans 6:23, we all fall short and need to be saved by God's gift through Jesus. The good news for all is the opportunity to receive the free gift, from God which was Jesus' death on the cross to pay for our sins, resulting in eternal life from the moment of acceptance to eternity in heaven. A helpful biblical passage for understanding the level of commitment that follows is Romans 10:9-10: "That if you confess with your mouth

'Jesus is Lord' and believe in your heart that God raised him from the dead, you will be saved. For it is with your heart that you believe and are justified, and it is with your mouth that you confess and are saved." One result of accepting His gift of salvation by giving your life to Him is that you will take seriously what it means for Jesus to be Lord or "boss" of your life to the point where others around you know and observe this.

Once this occurs, the direction of your life will change with the main focus being on God. Your heart, mentioned in this last passage, is sort of the compass for your life, pointing to where it needs to go. Once you have surrendered your heart to Christ, God's way is usually in the opposite direction of the way you were heading. The heart is important in this process since it is what truly reveals the sincerity of your commitment.

There is one last Bible passage I would like to share for it provides a good overview of what a Christian conversion entails. It is Ephesians 2:8-10 which states,

> For it is by grace you have been saved, through faith– and this is not from yourselves, it is the gift of God– not by works, so that no one can boast. For we are God's workmanship, created in Christ Jesus to do good works, which God prepared in advance for us to do.

Grace is defined as unmerited favor, meaning even though we do not deserve being saved from the punishment from our collective sins, God does it out of His great mercy and love for us. Because God alone does it, we cannot boast but rather all the credit and glory goes to God. Once we are His, we become His personal workmanship, as He shapes and develops us to do the work He has planned for us to do. He has a plan for your life that involves using the gifts and talents He has given to you to carry out the work He planned in advance.

Imagine God's plan involving every single Christian working on His master plan, flowing as a team made up of people with the appropriate gifts that complement each other as we all work on His plan. That is what the Church is to be, not a building but rather a collection of believers joined together as the spiritual family of God to function as His working body on Earth.

Where is your life at this stage in your journey? Do you feel rudderless, without a sense of direction? Is there a lack of peace and purpose in your life? Are you uncertain about the future and in particular about your future after life on Earth? Is there a lack of joy in your life?

Maybe when you were younger you strayed from God, or maybe you did not even realize that to become a Christian takes more than being involved in church. As a follow up to a *God Discovery Banquet* presentation, my wife and I went over a study called *Christianity Explained* with a couple who wanted to know more. The first course sessions lay out the foundation of the Christian faith. Toward the end of the course, once the attendees understand what it means to be a Christian, they are asked if they want to make a commitment to Jesus.

At this point, what the husband said was quite enlightening. He said that he did not even realize it took a personal lifelong verbal and heartfelt commitment to become a Christian. I suspect there are many who have this same misunderstanding. Immediately he said he wanted to give his life to Jesus, with his wife also wanting to make a renewed commitment.

To actually take that step of accepting Jesus' death on the cross for your sins and then surrendering your life to Christ, you need to have a basic conversation with God, one on one, stating your desires to do so. You can talk or pray directly to God for He will hear and listen to you. An example of the prayer you can pray can be the following. (You can elaborate and personalize each phrase to fit who you are):

> Oh God I know I have fallen short of what I was meant to be in Your eyes. I have come to realize You are a perfect and holy God but thankfully a merciful and all-loving God. You cannot allow any sin or actions done in disobedience to You to go without being addressed. Again, I now know I have fallen short because the whole focus of my life has been me and my selfish desires to run my life my way. I ask for Your

forgiveness for all my sins. I thank You that You are reaching out to me in love, a love that was ultimately displayed through Jesus dying on the cross to take the punishment for all my sins. I humbly accept that free gift of Your death on the cross and in appreciation yield the total being of my life to you. I ask that You take my life and begin working on it to make me a new person according to the purposes You have for me. I want to be Your workmanship. I thank You for giving me eternal life with You which begins now, and I look forward to having You be the Lord and Master of my life. Amen.

If you sincerely prayed a prayer like the one above, you need to do a few things so you can start to grow your personal relationship with God. To cultivate your relationship with God, you must begin to read the Bible on a daily basis. Obtain a good study Bible that provides background and explanations of the texts you read. Write on the inside cover a note describing your commitment and the date you made it and then sign it as a reminder of the time you made that decision for God.

The place to begin reading in the Bible are the first four books of the New Testament: Matthew, Mark, Luke and John. Those four gospels tell about Jesus' life and ministry while He was on earth in bodily form. As soon as possible, tell another committed believer about your confession of faith in God. Seek a Christian who has been a believer for a few years who can come along side you to answer questions, pray with you, and support you in your new life.

Part of this will be participating in a Bible-believing church and will include attending worship services and Bible studies and/or growth groups, and serving in the church according to the talents and gifts God has given to you. Sometime soon you should consider the church baptizing you, which is a public proclamation of the new life you are beginning in Christ. I would love to hear from you if you did take this life-changing step of faith. I can be contacted by email at GodDiscoveryBanquet@gmail.com.

I pray this book has made you think about God and opened your eyes in wonder at His work and creation. I further hope it has brought you to the realization that the Bible is the only true source of information about God and as a result has given you a lifelong passion to read and re-read it in order to gain a better understanding of God, which will lead to a deeper personal relationship with Him. My final desire is that this book has helped set your life on a course that is intimately intertwined with God and His plans for you. May the Lord richly bless you in this endeavor.

Appendix One

Here are the 67 Messianic prophecies concerning Jesus I mentioned in chapter 14.

Retrieved from http://www.jesus-is-savior.com/Miscellaneous/messianic_prophecies.htm

Prophecy Testament	New Testament Reference	Old Testament Reference
His pre-existence	Micah 5:2	John 1:1, 14
Born of the seed of a woman	Genesis 3:15	Matthew 1:18
Of the seed of Abraham	Genesis 12:3	Matthew 1:1-16
All nations blessed by Abraham's seed	Genesis 12:3	Matthew 8:5, 10
God would provide Himself a Lamb as an offering	Genesis 22:8	John 1:29
From the tribe of Judah	Genesis 49:10	Matthew 1:1-3
Heir to the throne of David	Isaiah 9:6-7	Matthew 1:1
Called "The mighty God, The everlasting Father"	Isaiah 9:6	Matthew 1:23
Born in Bethlehem	Micah 5:2	Matthew 2:1
Born of a virgin	Isaiah 7:14	Matthew 1:18
His name called Immanuel, "God with us"	Isaiah 7:14	Matthew 1:23
Declared to be the Son of God	Psalm 2:7	Matthew 3:17
His messenger before Him in spirit of Elijah	Malachi 4:5-6	Luke 1:17
Preceded by a messenger to prepare His way	Malachi 3:1	Matthew 11:7-11
Messenger crying "Prepare ye the way of the Lord"	Isaiah 40:3	Matthew 3:3

Would be a Prophet of the children of Israel	Deuteronomy 18:15	Matthew 2:15
Called out of Egypt	Hosea 11:1	Matthew 2:15
Slaughter of the children	Jeremiah 31:15	Matthew 2:18
Would be a Nazarene	Judges 13:5 Amos 2:11	Matthew 2:23
Brought light to Zabulon and Nephthalim, Galilee of the Gentiles	Isaiah 9:1-2	Matthew 4:15
Presented with gifts	Psalm 72:10	Matthew 2:1, 11
Rejected by His own	Isaiah 53:3	Matthew 21:42 Mark 8:31, 12:10 Luke 9:22, 17:25
He is the stone which the builders rejected which became the headstone	Psalm 118:22-23 Isaiah 28:16	Matthew 21:42 1 Peter 2:7
A stone of stumbling to Israel	Isaiah 8:14-15	1 Peter 2:8
He entered Jerusalem as a king riding on a donkey	Zechariah 9:9	Matthew 21:5
Betrayed by a friend	Psalms 41:9	John 13:21
Sold for 30 pieces of silver	Zechariah 11:12	Matthew 26:15 Luke 22:5
The 30 pieces of silver given for the potter's field	Zechariah 11:12	Matthew 27:9-10
The 30 pieces of silver thrown in the temple	Zechariah 11:13	Matthew 27:5
Forsaken by His disciples	Zechariah 13:7	Matthew 26:56
Accused by false witnesses	Psalm 35:11	Matthew 26:60
Silent to accusations	Isaiah 53:7	Matthew 27:14
Heal blind/deaf/lame/dumb	Isaiah 35:5-6 Isaiah 29:18	Matthew 11:5
Preached to the poor/brokenhearted/captives	Isaiah 61:1	Matthew 11:5
Came to bring a sword, not peace	Micah 7:6	Matthew 10:34-35
He bore our sickness	Isaiah 53:4	Matthew 8:16-17
Spat upon, smitten and scourged	Isaiah 50:6, 53:5	Matthew 27:26, 30

Smitten on the cheek	Micah 5:1	Matthew 27:30
Hated without a cause	Psalm 35:19	Matthew 27:23
The sacrificial lamb	Isaiah 53:5	John 1:29
Given for a covenant	Isaiah 42:6 Jeremiah 31:31-34	Romans 11:27/ Galatians 3:17, 4:24 Hebrews 8:6, 8, 10; 10:16, 29; 12:24; 13:20
Would not strive or cry	Isaiah 42:2-3	Mark 7:36
People would hear not and see not	Isaiah 6:9-10	Matthew 13:14-15
People trust in traditions of men	Isaiah 29:13	Matthew 15:9
People give God lip service	Isaiah 29:13	Matthew 15:8
God delights in Him	Isaiah 42:1	Matthew 3:17, 17:5
Wounded for our sins	Isaiah 53:5	John 6:51
He bore the sins of many	Isaiah 53:10-12	Mark 10:45
Messiah not killed for Himself	Daniel 9:26	Matthew 20:28
Gentiles flock to Him	Isaiah 55:5, 60:3, 65:1; Malachi 1:11; II Samuel 22:44-45; Psalm 2:7-8	Matthew 8:10
Crucified with criminals	Isaiah 53:12	Matthew 27:35
His body was pierced	Zechariah 12:10 Psalm 22:16	John 20:25, 27
Thirsty during execution	Psalm 22:15	John 19:28
Given vinegar and gall for thirst	Psalm 69:21	Matthew 27:34
Soldiers gambled for his garment	Psalm 22:18	Matthew 27:35
People mocked, "He trusted in God, let Him deliver him!"	Psalm 22:7-8	Matthew 27:43
People sat there looking at Him	Psalm 22:17	Matthew 27:36
Cried, "My God, my God why hast thou forsaken me?"	Psalm 22:1	Matthew 27:46
Darkness over the land	Amos 8:9	Matthew 27:45
No bones broken	Psalm 34:20 Numbers 9:12	John 19:33-36
Side pierced	Zechariah 12:10	John 19:34

Buried with the rich	Isaiah 53:9	Matthew 27:57, 60
Resurrected from the dead	Psalm 16:10-11; 49:15	Mark 16:6
Priest after the order of Melchizedek	Psalm 110:4	Hebrews 5:5-6; 6:20; 7:15-17
Ascended to right hand of God	Psalm 68:18	Luke 24:51
LORD said unto Him, "Sit thou at my right hand, until I make thine enemies thy footstool"	Psalm 110:1	Matt 22:44; Mark 12:36; 16:19; Luke 20:42-43; Acts 2:34-35; Hebrews 1:13
His coming glory	Malachi 3:2-3	Luke 3:17

Appendix Two

The table was taken directly from the reference website as we discussed in Chapter 14. The table in this Appendix lists 67 different prophecies that were fulfilled. You can read all of the stated prophecies in the table, but for the sake of brevity, I will briefly look at only 12 of the ones that may be the most familiar to all.

For those who celebrate Christmas, a holiday established to remember and celebrate the birth of Jesus, the following should be familiar:

1. born in Bethlehem,
2. born of a virgin,
3. His name called Immanuel, "God with us," and
4. presented with gifts. These events and details that surrounded Jesus's birth can be readily found in Christmas carols, Handel's *Messiah,* and at any candlelight Christmas Eve service.

At the other end of Jesus's life, the following prophecies clearly foretold the events related to His death and the week leading up to his crucifixion (keep in mind, no ordinary person could have preplanned these events such as those relating to His birth and death, thus making Himself the Messiah—unless He really was).

1. He entered Jerusalem as a king riding on a donkey, which is exactly what happened on Jesus' entrance to Jerusalem on Palm Sunday;
2. He was betrayed by Judas, his associate, sold for 30 pieces of silver;
3. He was crucified with criminals as depicted by

any artist's rendition of the crucifixion which show three crosses with Jesus in the center and the two criminals on either side;
4. His body was pierced which refers to soldiers piercing Jesus's side with a spear to make sure he had indeed died;
5. Soldiers gambled for His garment, which they did while he was dying on the cross;
6. He was buried with the rich in a rich man's donated tomb;
7. He was raised from the dead, which is the focal event of Easter; and finally,
8. He healed the blind, deaf, lame and dumb, the hallmarks of Jesus' three year ministry.

These are only 12 of the 67 Messianic prophecies listed in this table, which is amazing considering the original writings of these prophecies occurred between 500 to 1000 years prior to Jesus' death. What is even more fantastic is that these 67 prophecies are not the complete list. If you go to the website that lists these messianic prophecies, you will find an additional 314 prophecies.

I know there will be those who read these and still be skeptical. Some will say, "The Old Testament prophecies had to be added or edited after Jesus life to fit the actual events of His time on earth." This is not possible when considering the Dead Sea Scrolls that were described in the archaeological section earlier. You will recall that those scrolls were written a few hundred years before Jesus came to earth and were copies of the books that make up the Old Testament. When comparing the Dead Sea Scrolls to the Old Testament we have today, they are virtually identical. That is indisputable proof that the Old Testament was *not* altered after Jesus' life on earth to make it appear that Jesus' life and death were predicted centuries earlier.

The other possible argument of skeptics is that what was written in the New Testament about Jesus did not actually happen but instead was written to match the prophecies of the Old

Testament. There is also a problem with this argument. From the textual integrity section above, you will remember the original copies of the New Testament were written within a few decades after Jesus' death, meaning there would have been an ample number of living witnesses to refute the writings—which did not happen. In addition, there are the ancient writings such as by the Jewish historian Josephus that corroborate the events described in the New Testament concerning Jesus. The only explanation then is that these Messianic prophecies were actually a supernatural act orchestrated by God not only to provide hope and a glimpse of the future for His pre-Jesus followers but also to give future followers the validation that the Bible is God's revealed and recorded Word, not someone's idea and personal thoughts about God.

Sources and References for All Figures and Tables

FIGURES

Figure 1. Picture taken by author (with a handheld box camera)

Figure 2. Salt and pepper picture found at Functional Earl salt and pepper shaker product for sale at trendhunter.com, Finance photo found on Shutterstock (used with permission).

Figure 3. Graph designed by the author

Figure 4. Taken then edited and improved by the author from https://en.wikipedia.org/wiki/File:Miller-Urey_experiment-en.svg

Figure 5. Taken from https://www.exeter.ox.ac.uk/morris-and-burne-jones-tapestry-to-be-housed-in-museum-quality-display-conditions/

Figure 6. Taken from https://en.wikipedia.org/wiki/DNA

Figure 7. (The Conversation, 2013)

Figure 8. Taken from http://upload.wikimedia.org/wikipedia/commons/6/6d/US_Navy_030529-N-5362A-001_A_U.S._Marine_Corps_Humvee_vehicle_drives_down_a_road_at_the_foot_of_Saddam_Hussein%27s_former_Summer_palace_with_ruins_of_ancient_Babylon_in_the_background. (*The appearance of U.S. Department of Defense (DoD) this photo does not imply or constitute DoD endorsement.*)

TABLES

Table 1. www.currentresults.com/Environment-Facts/Plants-Animals/estimate-of-worlds-total-number-of-species.php provides this compiled list.

Table 2. Compiled by author

Table 3. Information from *When Skeptics Ask* (Giesler, 2013, p. 222)

Table 4. Information from *A Ready Defense* (McDowell, p. 45)

Table 5. Information from *A Ready Defense* (McDowell, p. 45)

REFERENCES

The Conversation. (2013, June 24). *Kinky genes: how we fit three meters of DNA into a cell nucleus.* Retrieved from https://theconversation.com/kinky-genes-how-we-fit-three-metres-of-dna-into-a-cell-nucleus-15421.

Douglas, J. D. (ed); Hillyer, N. (ed); Bruce, F.F. (ed); Guthrie, D. (ed); (1962). *New bible dictionary.* Leicester, England: Inter-Varsity Press.

Fetherlin, B. (2013, December 1). Visions of Jesus. Alliance Life Magazine. Retrieved from https://www.cmalliance.org/alife/visions-of-jesus/

Geisler, N. L. (2013). *When skeptics ask.* Grand Rapids: Baker Books.

Ho, B., Baryshnikov, A., & Brown, G. W. (2018, February). Unification of Protein Abundance Data-sets Yields a Quantitative Saccharomyces cerevisiae Proteome. *Cell Systems*, 6(2), 192-205. DOI:10.1016/j.cells 2017.12.004.

Johnson, Todd. (2001). *Todd M. World christian encyclopedia.* Edinburgh University Press.

Kennedy, D. J. (1999). *Why I believe.* Nashville, TN: Thomas Nelson.

Kennedy, J. (1994). *What if Jesus Had Never Been Born.* Nashville: Thomas Nelson Inc.

Mahoney, T., Law, S. & Schroeder, G. (2015). *Patterns of evidence: Exodus.* USA: Thinking Man Media.

McDowell, J. (1979). *A Ready Defense.* San Bernardino: Here's Life Publishers Inc.

McDowell, J. (2009). *More than a carpenter.* USA: Tyndale Momentum.

McDowell, J. & McDowell, S. (2017). *Evidence that demands a verdict: Life-Changing truth for a skeptical world.* USA: Thomas Nelson.

Omenn, G. S, Lane, L., E.K., Beavis, R.C, Overall, C.M., & Deutsch, E. W., (2016). "Metrics for the Human Proteome Project 2016: Progress on identifying and characterizing the human proteome, including post-translational modifications." *Journal of Proteome Research*, 15:3951-3960.

Peet, J. H. *The Miller Experiment,* retrieved from *www.truthinscience.org.uk/content.cfm?id=3161*

Peretti, F. E. (2002). *This present darkness, Piercing the darkness, The Prophet.* [Set of 3]. USA: Mass Market Paperback.

Rawlings, M. *To hell and back*. Nashville, TN: Thomas Nelson.

Stroble, L. (2016). *The Case for Christ: A Journalist's Personal Investigation of the Evidence for Jesus*. USA: Zondervan.

Stroble, L. (2014). *The Case for a Creator: A Journalist investigates Scientific evidence that points toward God*. USA: Zondervan.

Smith, Kevin. God Discovery. *God Discovery Banquet* [DVD]. Bluefish. Retrieved from https://www.goddiscoverybanquet.com/

Thaxton, C. B.; Bradley, W. L. & Olson, R. L. (1984). *The Mystery of Life's Origin*. USA: Philosophical Library Inc.

Vander Laan, R. (Director). (2018). *That the world will know* [DVD]. USA: Zondervan.

Wilson, J. (2002). *Molecular Biology of the Cell*. 4th Edition. New York, Garland Publishing.

Wikipedia. (2021). *Miller-Urey experiment*. Retrieved from https://en.wikipedia.org/wiki/File:Miller-Urey_experiment-en.svg

About the author, Dr. Kevin Smith, and the God Discovery Banquet

Kevin has lived in Western Pennsylvania all his life and grew up on a farm outside of Smithton, PA. When he was nine years old, during his attendance at a summer church camp at Camp Pine Springs in Jennerstown, PA, he made his first commitment to Christ. He graduated from South Huntingdon High School and attended Westminster College where he received a bachelor of science degree in chemistry and physics. The year after graduation, he taught high school science and chemistry at Rimersburg, PA, near Clarion, PA.

Next, he attended graduate school at the University of Pittsburgh chemistry department where he received his doctorate degree. He was a deacon and later an elder at Bellefield Presbyterian Church, which is located in the heart of Pittsburgh in Oakland. Bellefield has a large ministry to university students, international students, and the surrounding community. While attending there, Kevin was able participate in each of those areas. He also met and married the love of his life, Janet, at Bellefield. They have three grown children, Rachel, Sharon and Ben, and four grandchildren, Rylan, Jude, Arlo, and Lucy.

During his graduate and postdoctoral work, he published more than a dozen scientific journal articles. He worked for the Advanced Materials Corporation as a research and development scientist, serving as general manager of the Hydrid Division. Next, Janet and Kevin moved to Latrobe in 1990 where he taught general chemistry and physical chemistry at Seton Hill and St. Vincent Colleges for one year. For ten years, he was the senior technical analyst at NASA's technology transfer center at the University of Pittsburgh where he assisted in the transfer to U.S. companies of technologies that were developed at NASA, DOD, DOE and Department of Agriculture laboratories.

Kevin has four patents and for 15 years was involved as a technology consultant and a business owner of technology-based companies. The businesses were related to consulting for technology transfer activities, a business that assembled and fabricated 80% of a 98% efficient permanent magnet motor for marine use, researched and developed nanomaterials for use in antimicrobial and mold resistant applications for use in medical, construction and aquaculture applications, and the development of metalliding of metallic surfaces to make them hard and corrosion resistant. He served as an elder at the Latrobe Christian and Missionary Alliance Church for twenty years.

For the past 45 years, he has studied the Bible and the Christian faith in parallel with his scientific studies and endeavors and has been amazed at how science and other evidence support and provide proof that there is a God and that the Bible reveals the truth of who God is. He is excited to share in this book just a small glimpse of this evidence in the hope that it will be a blessing to you and will encourage you to have a relationship with God.

Made in the USA
Middletown, DE
04 July 2023